MEN MAY COME
and
MEN MAY GO,

BUT I'VE STILL GOT MY

LITTLE

Pink

RAINCOAT

ALSO BY
Gigi Anders

Be Pretty, Get Married, and Always Drink TaB:
A Memoir
(formerly *Jubana!*)

Gigi Anders

rayo *An Imprint of HarperCollinsPublishers*

MEN MAY COME

and

MEN MAY GO,

BUT I'VE STILL GOT MY

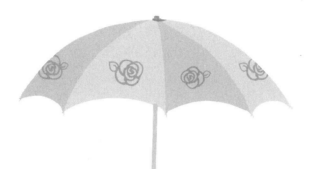

LITTLE

Pink

RAINCOAT

Life and Love In and Out of My Wardrobe

HarperCollins books may be purchased for educational, business, or
sales promotional use. For information, please write: Special Markets
Department, HarperCollins Publishers, 10 East 53rd Street, New York,
NY 10022.

FIRST EDITION

Designed by Janet M. Evans-Scanlon

Library of Congress Cataloging-in-Publication Data has been applied for.

ISBN: 978-0-06-111885-2
ISBN-10: 0-06-111885-0

07 08 09 10 11 DIX/RRD 10 9 8 7 6 5 4 3 2 1

AUTHOR'S NOTE

I love a good suit. Pant or skirt or bathing. But not a libel suit. With the exception of four men's real names, you won't be reading any other real men's real names in this nonfiction book. I've also changed or omitted some of the particulars to protect people's privacy.

—*G. A.*

To Billy,

For never telling me what not to wear or who not to sleep with.

I thank you and I love you.

CONTENTS

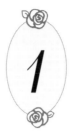

Little Pink Raincoat

Suddenly, on the second Sunday of March in 2003, it appeared. My "it" hit me like a *coup de foudre*, a French lightning bolt. Only it was American. I was zapped by the picture of a model in the "Sunday Styles" of the *New York Times*, wearing the Gap's little pink raincoat. Not just *any* little pink raincoat. No. That would be plebeian. MY little pink raincoat was the *cutest*, most perfect, most I've-got-to-have-you-this-very-second-or-I-will-die-lonely-and-raincoatless little pink raincoat. That's right. That's how good it was. It was my rainy day destiny.

Men move me, but not like clothes and accessories do. Or maybe they're inextricably bound. Even saying "clothes and accessories," uch, that sounds so ordinary and banal. Can any kiss, any flower, any orgasm—approximate the rush of finding That Perfect Pink Raincoat? I mean the world's grooviest raincoat. I mean the raincoat of laserlike focus and obsessive desire. The raincoat that stops you cold. The elusive one that once you have it, you feel not just better, but fulfilled, transformed, you've arrived more deeply and forever at your best self.

Do you understand what I'm talking about? Of course you do. Something that makes you feel scared thinking that if the *Times* just hadn't come that one day, you would have missed this. You'd have lost your only chance at THE little pink raincoat of all time. (It's fun to scare yourself with contemplative what-ifs, but not until you've landed the damn thing and you know no one can ever take it away from you.)

Here she was: Simple. A-line. Hidden buttons. Bright but not too bright pink. Soft but not forgettably soft pink. Baby girl pink. Dubble Bubble pink. Daisy Buchanan pink. Billowy clouds streaking across the sky at twilight pink. Underneath it, the brunette model was wearing a man's white cotton long-sleeved shirt over broken-in, slightly faded skinny blue jeans. The picture was cropped there, so I kind of wondered about the shoe situation. But not for long. I had to get on the phone and order my love dream to be sent to my house so I could go on living in love and dreams. That was the subtext behind the impulse, it's what's always at the hopeful, beautiful, beating heart of it: Love. And the vertebrate with whom I was engaged in beasty love, he would HAVE to marry me once he saw me in that little pink raincoat. He would HAVE to. He couldn't not. It was too lovely. And sweet. Innocent, almost. Girlie, but not froufrou. Just really, really, really great. I was thinking how I'd pop out in a crowd of boring beige and plaid Burberrys, like a lone little pink flower in a desert.

I had my *idée fixe* all worked out: Little white cap-sleeved T-shirt, white lace-edged bra (so you could see it, but almost, like, accidentally), fitted black cotton capri pants, black leather ballet flats with quilted black patent leather tips across the toe box. Legs ultra-shaved and self-tanned. Chanel Nº19, for sure—it was almost spring. Black kohl liner the French way, on the inside of my eyelids, and tons of black mascara, but really worked through so there's no *hint* of glop or flake. Love that look. Soft pink lipstick, maybe just a lot of pinky gloss. Maybe like NARS's "Orgasm" gloss but pinkier, more full-bodied. Then ice-white acorn pearl drop earrings with the silver hooks and my watch with the black faux-croc band and silver and white face, and my Isaac Mizrahi for *Tarjay* black leather purse with the silver zipper.

And, like the wedding cake toppers to top all such toppers—the little pink raincoat. A vaguely Gallically gamine ensemble that I'd still be proud to wear twenty years from now. That's the secret. Always ask yourself: "How mortified would I be if I saw myself in this outfit, say, post-menopause?"

Done.

After dating and living together off and on for four years, The Dinosaur—he was, after all, twenty years my senior—still would not commit. We'd even been engaged at one point, for about three or four minutes, and he'd broken *that* off. When he did, I thought my life was over. So I plunged into the requisite Madame Butterfly mode—I've always been a little dramatic. This involved crashing into the tragedy of love gone wrong twenty-first-century style, on my aging mattress that was turning into a hammock no matter how many times I turned and flipped it, and armed with Parliaments and Ben & Jerry's New York Super Fudge Chunk. In a twisted way, this is the fun part. When they leave you, you get to eat whatever you want, as much as you want of it, turn into a total slob of whimpering self-pity, and check the fuck out. It's like getting a guilt-free temporary pass from life and its grown-up responsibilities. You get to Really Suffer and Do It Up Right.

But then—this is the bad part—those who dumped you can

come to their senses and come back at an incredibly inconvenient time. At least this one did: After I'd gained fourteen pounds and my face was all broken out and I was at my haggiest, most reclusive, most kill-me-now worst. Yes, The Dinosaur returned. Despite the fact that the only thing that fit me was my wardrobe of black yoga pants (J. Crew's and Old Navy's are the best), I was actually happy about this. When you're crazy in love and crushed, you're dying to get uncrushed ASAP. In this state, the man you love, and only he, can make it better. So I tearfully and gratefully took The Dinosaur back. The pain and the Ben & Jerry's bingeing stopped. Or maybe they just went underground. Either way, I got my love back and I got back into my size 8 jeans.

The Dinosaur and I rejoiced on my sagging mattress. Afterward, we shared my sensational sesame noodles. This is not a metaphor. I make the world's best sesame noodles, if I do say so myself. One of The Dinosaur's predecessors, a Brooklyn mama's boy in his forties who'd never not lived with his mother (which is why he became my ex), once ate so many of my famous sesame noodles at a single sitting that he spent the subsequent twenty-four hours locked in my powder room, emitting truly terrifying noises and odors. I took it as a kind of compliment.

Tip: What makes a recipe good enough to make a person sick is the same secret as that of the flawless red lipstick shade—you have to combine at least two. In the noodles' case—leave it to Jews and Italians to improve an Asian dish—it was Arthur Schwartz's *What to Cook When You Think There's Nothing in the House to Eat* and *The New Basics Cookbook* by Julee Rosso and Sheila Lukins. As for The Dino's and my post-noodles dessert, we compromised and had coconut sorbet with fresh strawberries. I personally hate healthy desserts—what's the oxymoronic point?—but The Dinosaur's always trying to eat "lite." At least the coconut was bad for us. If it had been 2005 instead of 2003, Ben & Jerry would have already invented Fossil Fuel, the sweet cream ice cream with chocolate cookie pieces, fudge dinosaurs, and a fudge swirl. I'm sure I could've forced that on my Dino and we'd both have been more fulfilled and stayed together forever.

But this was not to be. Because a few weeks later, The Dinosaur left me. Then The Dinosaur came back. Then The Dinosaur left me again. Getting the gist of my love problem? My beau was Lucy with the football and I was Charlie Brown. But hey, nobody's perfect. Give a boy his learning curve. Otherwise, my reptile had all the right stuff: Smart when not brain-dead, sexy when not preoccupied, sensitive when not oblivious, funny when not miserable, kind when not clueless, attentive when not distracted, generous when not cheap. He brought me Starbucks and bagels in the mornings. He sent me Martha Stewart roses just because. He was nice to Lilly, my kitty cat—even after she upchucked in his suede bucks. (In Lilly's defense, The Dinosaur wasn't wearing them at the time.) He wasn't fanatical about sports. He was a museum-goer. He liked Joni Mitchell.

How many straight men do *you* know like that?

Accordingly, I hung in there, well past closing time. Hello, four years. I tend to give a man I like the benefit of the doubt. No sense in throwing out The Dino with the antediluvian bathwater, right? So I gave him time for his romantic process, or whatever, that he needed to go through. You have to be patient with dinosaurs. You hit them on the tail and it takes two weeks for the message to reach the head. Jurassic Park wasn't built in a day. I must really love a challenge—romantic, culinary, sartorial, what have you. I think of what poor, dead JFK said about going to the moon, that we're choosing to go there not because it's easy but because it's hard. Exactly, Jack. Hard is good. It's more interesting.

Besides, I'm a Sagittarius. We are the zodiac's eternally sunny, hopeful sign. JFK was a Gemini, also a good sign and almost as lively as mine.

The Dinosaur, in contrast, was the Virgin. Virgo. Ooo. Darkness. Nowhere as scary as Scorpios, but still. The Dinosaur was, as is a typical Virgin Dino's wont, heavy. A brooder. At first, this seemed okay. Opposites attract. He grounded me, I lifted him up. But over time—did I mention it was four years?—The Dinosaur stayed slow. Real slow. Too slow. Slow to get it. "It" being not the fact of my high-voltage appeal—he got that fine—but of my suitability to be his bride. Why? Why is it that what attracts them is the same thing that keeps them at bay? I've never understood this. I'm not a violent person by nature. But sometimes I'd get so frustrated thinking I'd wasted my time and money on all those *InStyle Weddings* and *Martha Stewart Living Weddings* that I'd just want to grab that Dinosaur and shake-shake-shake him.

What was it? Our age difference? Not for me; I've always been attracted to older men. Somebody has to love them. And like me, this one was a Jew. Actually, he was less of a Jew than I was—he and the shiksa ex-wife always had Christmas trees, something I could no sooner do than wear black lipstick.

"You've never been married and divorced," The Dinosaur would always say, usually following a Manhattan or two. A *Manhattan*. Could there be a queerer cocktail? Those little maraschino cherries alone. Anyway, that was The Dinosaur's exclusive rationale. I wasn't a member of The Club, therefore I was ineligible to enter it.

"You don't know what it's like," he'd say.

"And your point is?" I'd say.

"That you don't know what it's like," he'd reiterate, chomping the cherry off its stem.

You don't know what it's liiike, baby, you don't know what it's liiike, *to love somebody, to love somebody, to love somebody the way I love you.* Thank you, Bee Gees. My relationship wasn't exactly turning out to be the love song I'd originally envisioned.

The Dinosaur would moan, "I may be terminally single." I assumed he'd picked that up from an old *Cosmo* in his shrink's waiting room. *Terminally single.* Please. It's not a disease! And tough, seasoned newsmen of a certain age like The Dinosaur—we'd met at a journalism conference back when I was a reporter—don't use those phrases, unless they're being sarcastic. Maybe my prehistoric boyfriend was trying to sound like a hipster. Eek.

My therapist, Manny, only keeps *National Geographic* in his waiting room. They do stories featuring bare breasts and the Cretaceous period, two of The Dinosaur's favorite things. About love and marriage, however, Manny called The Dinosaur "eternally ambivalent." This was not, obviously, what I wanted to hear.

"Are you saying it's time to release The Dino?" I asked him.

"There are many other fish in the sea," Manny said.

"But he's neither fish nor fowl. Of course, there are sea dinosaurs. Well, there were."

"Okay, well, you might try a mammal next time out."

Next time out? Was he kidding? I was exhausted. Four *years.* I didn't want a mammal, I wanted to fix the reptile I already had. *InStyle. Martha.* You know. Work with what you've got. Make a living fossil wreath. A dinosaur egg mobile. A scaly plate centerpiece. Something.

I poured myself a TaB (the can is pink, it gave me hope) and hit Gap's Web site. They had the coat. They called it a macintosh.

But they only had it in "antique white," "stone," and black. No, no, and no. I called 1-800-GAPSTYLE. My little pink raincoat was all sold out! Oh nooo! I thought, I'll bet super-acquisitive North-eastern chicks who didn't call the Gap's toll-free number would probably try calling individual Northeastern Gaps, but not with *my* singular zeal. I was mistaken. My little pink raincoat was sold out in my state, New Jersey, as well as in every New York and Connecticut Gap. That's right, I called every last Gap in every last city in those places. Well, I was gonna beat these dames at their own game. Nobody was gonna prevent me from MY marriage, dammit. So I devised a strategy. I proceeded to maniacally call each state in the United States, but alphabetically backward. So I began in Wyoming, where there is one solitary Gap. Hey, don't laugh. You never know.

"Are you calling about that mythical pink raincoat?" the lady asked. "I've heard about it, but we never got it. I guess Wyoming's not a pink raincoat kind of place."

Then, just for the hell of it, I tried North Carolina. I know, it wasn't alphabetically correct. But I used to live there. *Twenty-nine* Gaps later, *nada*. One woman laughed at me: "Tha-yutt pi-yunk raincoat? Oh honey, you couldn't fahnd tha-yutt on God's green earth now!"

This is my quest, to follow that pink raincoat, no matter how hope-less, no matter how far . . .

Back to the backward alphabet strategy. Wisconsin has nineteen Gaps. Luck struck on my very first city, a place called Green-dale. Was my impossible dream possible?

"Got it!" the young woman said. "Got the last medium right here. And boy, are you lucky. This thing's blown out of the store."

I whipped out my abused Mr. Visa and we charge-sent me my dreamy creamy pinksicle coat. Maggie Prescott would have been so proud. She was the imperious fashion editrix in the 1957 movie *Funny Face*, the one modeled after Diana Vreeland (who in real life said, "Pink is the navy blue of India"). Maggie Prescott sang "Think Pink!": . . . *Red is dead, blue is through, green's obscene, brown's taboo . . . pink's for the lady with* joie de vivre.

I have *joie*, Maggie. *Beaucoup joie.* So much *joie* that my latent hoarding gene—HNA, not DNA—kicked in. Maybe I needed *more* than one pink raincoat. What if the medium was too small? Or what if I had a spilling accident and needed a backup? No backup to be had at the Madison, Wisconsin, Gap, but Nancy, the store manager, promised to do a national search and get back to me in the morning. I didn't trust her.

"I swear I'll do it," she said. "It's a really cute coat. I want you to have it."

"Sure you swear," I told her. "But do you CARE, Nancy?"

"I do!" she said. "I swear I care!"

Having exhausted the remaining seventeen Wisconsin Gaps, I moved on to West Virginia, where there are four stores. They all laughed at me. One Gap guy, in Barboursville, said, "I was unpacking these last week or the week before and I thought, Who the hell's gonna wanna buy a Pepto-Bismol coat?"

"It's NOT Pepto-Bismol," I told him, even though I'd never even seen it.

"Three days later," he continued, "it was gone. We've been getting calls from Connecticut Gaps begging us for any size. I think I got one left in an extra-small."

Was this man taunting me? An extra-small wouldn't fit my bulimic cat Lilly and she weighs eight pounds.

I finally scored a large in Bellingham, Washington. That figures. It must be near Seattle, land of rain, suicides, and raincoats. I popped open a fresh can of TaB to celebrate and contemplate my pink arrivals. It was 12:30 A.M., six and a half hours after my quest commenced, but I was so energized by my coup that I called my friend Joan in Manhattan to tell her about it.

"Have you been on eBay recently?" she asked.

"No, why?"

"You said a pink Gap macintosh, right? They've got thirty-two of 'em on there selling for $150."

"What?!? I can't believe it! Little pink raincoat scalpers! Who thinks of this?"

"What do you care?" Joan said. "You've got yours. Good for you. It does look really cute. I wonder how it looks in person. Maybe I could wear it."

Was she insinuating something? I didn't want her to get any piggy-backing ideas, so I quickly hung up.

In the morning, Nancy from the Madison Gap actually called me back. I was stunned.

"I found it!" she proudly crowed. "I'm having a large sent right to you. They had one last one left at the Chicago store on Michigan Avenue. I must've called thirty stores."

I didn't have the heart to tell her I'd already found a large in Bellingham. So we charged and sent me a second large. Something was happening. This phenom was bigger than me and my three little pink raincoats. This is how features journalists like me think: Why pink? Why pink now? Was it because Jennifer Garner wore my raincoat that February when she hosted *Saturday Night Live*? That's what her publicist Nicole King thought.

"Jennifer's a very Gap girl," she said. "The *SNL* costume design team chose that coat for her for a skit."

Did she get to keep MY coat?

"I think she did," King said. "She really loves it."

"All I can say is that this pink raincoat is like a preppy, retro look," said my friend Mindy, a Washington, D.C., wardrobe stylist. "Very sweet, optimistic, and friendly, like grosgrain ribbon. Very Kennebunkport, yet über-feminine."

Then my sartorial soul mate Sarah Jessica Parker should have one! She even modeled for the Gap, albeit After Pink Raincoat, so she might've missed it. Plus, SJP's perfume Lovely and its print ads are all cotton candy pinkety-pink-pink.

"The Gap macintosh pink is a cool blue-pink," Mindy said. "It's like the Sweet'n Low packet pink. I'd put SJP in something warmer and softer, like . . . like an apricot Cosmo color. That's so her."

There's no escaping it. Single urban gals—okay, my friends and *moi*—are still obsessed with and in withdrawal from *Sex and the City*. It's sad, I know, but what can you do? My gay columnist friend Billy wrote an entire piece devoted to this in the fall of 2005. It was triggered by the fact that one Sunday night at Michael Jordan's The Steak House N.Y.C. in Grand Central Terminal, I saw Big. As in Mister. As in the actor Chris Noth. That hunky hunk-a burnin' love was seated with a guy behind a large pillar, enjoying an overpriced steak and wearing a kelly green shirt. It was all too much for me. So I passed out. Big didn't notice. It was really dark in there and I blended into the dark carpeting in my favorite black skirt (four floaty, gossamer layers of tulle and cotton and deconstructed ruffles at the hip. Devastating. *Anthropologie.com*, I love thee). A maitre d' and Billy quickly dragged me away by the armpits. I came to, typically, just in time for dessert. Big was gone by then but it didn't matter. I'd had my Big fix. I'd been Bigged.

"In Fantasy Land Mr. Big would have come to your rescue and fanned your wan and delicate face to revive you," said my Manhattan author friend Elizabeth. "But he never really was that kind of guy, was he? I was just watching a *Sex* rerun yesterday, when he tells Carrie begrudgingly, 'I fucking love you, okay?' "

"I saw that one last night too!" I told her.

"I saw Chris Noth walking in the Village one day," Elizabeth added. "He looked like he was just coming from the gym. A little frumpy and not at all Mr. Big-ish. But I still love him."

But I still love him. The eternal feminine position. Are we all masochists? Do we give men way more credit than they deserve? Should we reserve our credit strictly for plastic purchases? On their record *Nurds,* The Roches sing, *This feminine position tripped up with reptile into that most feminine position too fat to turnstile.* I'm not really sure what the lyrics mean, but I always appreciate the reptilian reference. Almost as much as I appreciated my new little pink raincoat. I knew Maryellen Gordon, *Glamour* magazine's deputy style editor, would too.

"Pink conjures up frothy bows and cotton candy," Maryellen said. "But designers have wrangled it into these womanly, gorgeous, grown-up pinks. And people tend to dismiss the importance of color in outerwear. When was the last time you saw a fabulous, great, light, happy, insouciant coat and said, 'Wow, I have to have that'? It's, like, 'Maybe if I buy this it will be 70 degrees tomorrow and I'll feel less depressed about the war.' Pink captures the zeitgeist."

It's definitely the opposite of war. Earlier that month, a women's group called Code Pink protested the war in Iraq dressed in pink.

"Obviously, people are a little bit freaked out about the war and Bush," noted my friend Andrea Linett. She's *Lucky* magazine's creative director. "People are dying for something to make life less depressing—black won't give you any excitement—and pink does that. It's Audrey Hepburn princess, Gwyneth Paltrow at the Oscars, Cinderella, girlie and more innocent than red, but not namby-pamby. After an icky winter, this pink raincoat is the fashion equivalent of 'Here Comes the Sun.'"

Maryellen Gordon said that huge companies like the Gap take a gamble on color—and it is an educated gamble because they know their business—but "I'm sure that right now they're wishing they had ten thousand more in production."

Were they? I called Gap spokesperson Erica Archambault.

"We believed in pink and selected it several months ago to be a key part of our spring line," she said. "It was planned and it's been very popular. We thought people around the country would be craving a burst of color this spring. It sounds dorky, but I put on pink and it makes me feel more cheerful when I'm in a bad mood."

Emotion! That's it! Pink is about feelings. *Joie.*

"I'd raise to sainthood whoever decided to do that coat," said Leatrice Eiseman, director of the Pantone Color Institute in Carlstadt, New Jersey. She and her fellow color gurus there study human response to color. They can tell a year ahead of time what'll be the next big hue. Corporations such as Nordstrom, Puma, and, yes, Gap, pay Pantone big bucks for its color savvy. In fact, the coveted collector's-item pink raincoat is Pantone's 15-2216, or Sachet Pink.

"Pink is a return to your little girl baby blanket," Eiseman said. "It wraps itself around you and protects you from the elements—like a raincoat does. It's nurturing and romantic and a harbinger of springtime. I think the mood now, with the somberness and darkness of war and having had a very bad winter, is perfect for it. Pink expresses all of those inner feelings in that one color."

I knew it! I was so jazzed over my findings that I called Joan back to breathlessly report. I told her about buying the second size large and about Jennifer Garner. I told her about Code Pink and "Here Comes the Sun" and—

"Wait a second," Joan said, interrupting me. She always interrupts me. It's part of our repertoire. "Did you say you ordered . . . TWO larges?"

Baby Medium arrived a few days later, like a pink bundle of love from heaven. Even better, the horizontal tab inside the collar, the one you can use to hang the coat on a hook, was— brace yourself—white and pink gingham. Baby Larges, the twins, arrived later that week. They were both too big. I sold one to my friend Mindy, who said it was not only *her* "perfect blue-pink" but that it would also complement her newly decorated, very tiny studio apartment. The first time I saw the place, Mindy proudly asked, "Doesn't it make you want to have sex?"

"Well," I told her, studying the deeply mauve walls, the black bearskin throw rugs, and the thirty-seven Hummer SUV-sized red, fuchsia, purple, and scarlet velvet shams neatly staggered across her heavy cherry velvet bedspread, "it does resemble the inside of a vagina."

I returned the other large little pink raincoat to my local Hackensack Gap, where a crazed cluster of seven women (customers *and* salespeople) lunged and all but tore it out of my hands. I may as well have been a Spanish *matadora* waving a pink cape. I'm telling you, obtaining this raincoat made me feel so triumphant, I ran outside and did the Rocky victory dance, right there in front of Riverside Square mall, humming the Rocky theme: Dada-DAAA, dadaDAAA, dadaDAAA . . .

I have always believed and still believe that said raincoat, as well as a few other choice garments, will get the men I love to finally commit and live with me tastefully ever after. Since I believe that, then you probably do too—if you're a female. Right? If we can somehow find that impossible-to-get-ahold-of little pink raincoat, he'll want us for more than datin' and matin', and we'll finally be Tiffany-little-blue-box happy. Now, it doesn't always have to be a little pink raincoat; it could be, say, a peach silk-satin bra with a matching panty, or a backless black dress. It's whatever you want to wear that makes you feel extra-good, alluring, and refreshed. It's devastating, crazy, romantic, deluded, and profoundly feminine, this intense and magical power we ascribe to clothes, jewelry, underwear, scarves, shoes, sunglasses, makeup, and even to perfume, to get us closer not just to men and sex—you can get those anywhere, *quel* bore—but to love, the biggest, hardest, most elusive treasure of them all. That's the Little Pink Raincoat state of mind.

This, of course, never works—in the long run. Clothes may make the woman but clothes don't make the man stay with us. Not even the drop-dead cutest ones—clothes or men. (Who knows why men stay with us? I have no idea. I guess it's because they love us, preferably naked, therefore canceling our entire shopping therapy core belief system.) But! That doesn't mean we *care* that it never works. We *think* that it works, we *believe* that it works, we *hope* that it works, and that's what counts. (It's also what keeps every one of my fave stores, from Banana to Bergdorf, in business.) Besides, it really doesn't matter if men leave us. They do this. It happens. We cope. We forge ahead, plastic at the ready. Because guess what? After they leave us, we still get to keep the outfit. That's concrete. That's tangible. That's real. There are always other fish in the sea, as Manny said, but he's a *man*-ny and you may never see that orgasmic Chanel red lipstick again because they might discontinue it because they just introduced it for, like, one season or something. Chanel is fleeting and unique. You see what I mean? Lose the guy. Keep the lipstick. Life and love are so much less disappointing that way.

In that sense, my little pink raincoat was, literally, a little pink raincoat. But it was also a metaphor for feeling that *feeling* that a thing *like* a little pink raincoat gives. My beloved *Washington Post* Style editor, Peggy Hackman, agreed with me. So much so that I wrote a whole article about it the following month. The day my story "Forever Chasing Raincoats: Going for the Pink" ran, none other than the *Today* show's Katie Couric talked about it on the air. Katie said something like "I want this coat!" I must've hit a feminine nerve. I was on to something. If women go berserk over a little pink raincoat, then we must go berserk over things *like* it. We know what you wear can change your life. We also know it can*not* change your life. See? The point is we *feel* it can, and that kind of yearning for a changed life means a change toward love, both inward and outward.

May would be good for a wedding! The Dinosaur and I could honeymoon in Vernal, Utah. That's where Dinosaurland is. They have a HUGE pink Sauropod with flirtatiously long black eyelashes and a great big friendly Pepsodent smile welcoming you at the entrance. The Dinosaur could commune with his kin, or at least their fossils. Then we could dine at The Dinosaur Brew Haus ("Beer, Burgers, and BONES"), and grab a frappé or a latte nightcap at the nearby Shivers-N-Jitters café. It would be so groovy. Unless he breaks up with me first.

Which he did.

I remember this jock I knew in college. He was newly engaged. He was just a kid, barely twenty, and his zombified fiancée of choice seemed in dire need of a personality implant.

"I know she's not the prettiest," he told me one day. "And I know she's not the smartest. And I know she's not the richest."

"So why are you marrying her?" I asked.

"I'm sick of first dates."

Hey, try four years of the same date. But! My little pink raincoat would renew The Dino and me. It was spring, life was abloom, we could be too. As he lived near Philadelphia (we were weekend commuting when communing), I pictured him all alone, bereft and tragically trudging along the banks of the Schuylkill River or the Manayunk Canal. He'd be bitterly pondering the slogan "City of Brotherly Love." Trudge, trudge, trudge. Lonely cheese steak Dinobeest.

"Checked out any dating sites lately?" my New Jersey journalist friend Eman asked me one day over the phone while I was in full-throttle Madame Butterfly mode.

"What?" I said, confused. "No, why? I'm not ready to date other people. I love The Dinosaur! And I'd never go on one of those sites. I was watching Dr. Phil today and he doesn't believe in them. He goes, 'If your only standard is that the other person has to own a computer—' "

"Are you online right now?"

"No," I said, lighting a Parliament. "I'm too busy being suicidal on my hammock which used to be a bed, thanks."

"Your Dinosaur is on there," she said. "In living Dino color."

"What are you talking about?" I asked, nervously reaching for an ashtray and my New York Super Fudge Chunk. "My Dinosaur's melancholy and meditating gloomily down by the river, channeling Ben Franklin, or whatever."

"Well, I was curious so I did a little luuuv search on your allegedly tortured OCD Rain Man. Typed in his general peripherals, and bada-boom, baby. 'Advertisements for Myself.' Here, I'll send you a couple of the links."

"HE'S ON MORE THAN ONE?!?" I screamed.

"No river gloom action," Eman said. "He's just looking for *action.*"

I dropped my ice cream and the phone and bolted to my computer, the broken pieces of what used to be my unbroken whole heart banging madly. It couldn't be. It couldn't be MY Dinosaur on there. It had to be someone else's dinosaur. I clicked on one of the links and . . .

Oh. My. God.

Oh. My. Dinosaur.

Standing in front of the Chelsea Hotel. Smiling. SMILING. In his nerdy old eyeglasses. (Yes, I'd fallen in love with a man with Bad Glasses. *That's* true love.) With his beat-up old khaki trench coat draped over his forearm. We'd been in New York City visiting his sister and brother-in-law for Thanksgiving just four months earlier. And now he was using the picture I'd taken of him to get dates with other women!

JOURNALIST SEEKS THAT ELUSIVE COMBINATION OF ADVENTURE AND STABILITY

> Some of my friends sometimes refer to me as a dinosaur, mostly in terms of old values and doing things right because that's the way you do them.

Holy shit.

> I'm looking for someone who's very smart, down-to-earth, attractive and, above all, honest without being unkind. A good sense of humor is a requirement of survival with me.

Hello?!?

> I need my own space and respect my partner's,
> but I'm also looking for something more than a
> weekend date.

Liar, liar, pants on fire! You're eternally ambivalent!

> At this stage, there's no one around without bag-
> gage. You just look for matching luggage.

More shrink's-office women's-magazine twaddle! Was my would-be husband dragging his unmatched suitcases by the banks of the Manayunk, stopping occasionally to check his new *luuuv* *.com* e-mails on his BlackBerry? Jesus!

Turn-ons:

> Boldness/Assertiveness, Brainiacs, Flirting, Public
> displays of affection.

Turn-offs:

> Body piercings, Tattoos, Thrill seekers.

Breathless and tearful, I found my phone under three pillows and called Billy. At times like these, a girl needs a real homo to straighten her out.

"Basically what he's saying," Billy said, "is 'I want someone exactly like you. Except not you.' "

So I did what any literate, homicidal maniac who's been turned into romantic roadkill would do: I responded to the personal ad.

I am, after all, almost exactly like me.

Hi,

I'm a cute little Jewish *princesse* living in New Jersey. Really loved your ad and thought you might like to get together. I know this great little pizza place in Hackensack. Nothing fancy, but the pizza's orgasmic. I was going to go tomorrow for dinner with some-one else, but he's too unstable. If you're available for a highly ca-loric adventure, please let me know.

Elusively yours, *La Princesse*

P.S.: Loved your picture in front of the Chelsea Hotel. Did I take that or was it your brother-in-law? :)

I closed my eyes, took a deep breath, and hit SEND.

Thirty seconds later, my phone began ringing. I glanced at the caller ID. It wasn't one of my 500 creditors. No, it was The Dino. I let the call go to voice mail. I lit a Parliament and poured a TaB. My cell phone began ringing. Dino. I let that one go to voice mail, too. Hm-hmm-hmmm. *How does it feel? . . . You're invisible now, you got no secrets to conceal. How does it feel?* The Bee Gees, God love 'em, are no match for Bob Dylan. Nobody is. The regular phone began ringing again. Dino. I picked it up.

"Hell-ooo?" I chirped.

"Hi," he said severely. "So, um, huh. I got your little note here."

"Uh-huh."

"Is this . . . some kind of . . . joke?" he asked.

"Oh no," I said. "It's deadly serious. We're talking pizza."

"Oh. Okay. Well."

"Did you honestly think I wouldn't see your ad? Like, I'm so stupid? I spend my life on the Internet, what is WRONG with you?"

"You're anything but stupid."

"Oh, so you're saying you *wanted* me to see this."

"No. I don't . . . know. I wasn't thinking—"

I sipped my TaB and waited. Long silence. Tick. Tock. Tick. Tock. It was a scintillating conversation, all right. I know I was tingly all over. Tick . . .

"So, you wanna go get pizza, or what?" The Dinosaur finally sputtered.

"Um, sure," I said. "Since you put it that way. How could I resist?"

"I'll drive up tomorrow. We can do a late lunch-early dinner thing. I wanna drive back before it gets too dark."

Translation: "I'm not sleeping over so don't even think about it."

Yeah, but he hadn't seen me in my new little pink raincoat. I walked to the hallway closet and pulled it out. Gorgeous.

"It's you and me, sweets," I told it. "We're fixing him together."

I awoke the next morning all giddy. *Idée fixe*—actuated. Hair and makeup—perfect. Outfit to go under the little pink raincoat—perfect. Jewelry and purse—also perfect. I sprayed on some Chanel Nº19, slid on my raincoat—I couldn't wait—and looked in the full-length mirror. An extra dab of NARS's Orgasm gloss, and I'd be good enough to eat.

The Dinosaur arrived exactly at the appointed hour, four o'clock. Considering that I'd been ready since noon, I'd had plenty of time to calm down and break myself in to the pink fabulosity of it all. (The first time you wear something new, you're always a little self-conscious about it. It just takes that one time, and then you're fine. *You're* wearing *it* rather than vice versa.) The Dinosaur kissed me hello hurriedly. He was wearing a heavy down coat and a black woolen cap. It was probably still way too cold and unrainy outside for my springtime outfit, but I didn't mind. I can always suck it up for beauty and love and dreams.

"So you're all ready to go?" The Dinosaur asked.

"Let's go," I said, waiting to see how soon he'd swoon over my coat.

"Aren't you gonna be kinda cold in that thing?" was all he said.

The restaurant wasn't busy at this off-hour—4:10—and we were seated right away. I was freezing so I kept my prized coat on. Besides, this would prolong the pink enchantment. Oh, I just love this place, my favorite pizza place in all New Jersey, the incongruously named Brooklyn's Brick Oven Pizzeria. It's just down the street from the Hackensack Cemetery. I probably should've heeded the symbolism, relationship-wise.

The Dinosaur ordered for us: Red wine, salads, a large pizza. I lit a cigarette and felt very content. I love Jersey, if only because it still has smoking sections. In some ways, the entire state is a smoking section. A Latina waitress came over with glasses of ice water and a basket of warm garlic bread.

"*Muy bonito,*" she said, nodding at my coat. "Preety peenk."

"*¡Gracias!*" I said.

I looked at The Dinosaur. *Nada.* He looked moribund. How was it possible that a complete stranger—okay, granted, it's a woman, but still—can see me, and my own almost-husband cannot?

"How's tricks?" I asked him, tucking into the delicious greasy bread.

"Okay," he said, grimly sipping water.

"This is really good bread," I said. "Don't you want some?"

"Later," he said. Apparently we were going to have an exchange in which I'd supply the animation and he'd supply the suspension. Kind of like the history of our entire relationship.

The waitress brought our wine and salads.

"Cheers," The Dinosaur said, looking anything but cheerful. Maybe he'd be perkier next door, at the necropolis. Why hadn't he mentioned my coat? Why wasn't he talking to me? Why was he freakin' dating online?

The pizza arrived, fragrant and bubbling. This pizza rocks my world. I shook grated Parmesan on my half—extra cheese-mushroom-extra sauce-pepperoni—and The Dinosaur gnawed his fresh tomato-black olive-baked garlic cloves-fresh spinach. Like his Mesozoic-era terrestrial carnivorous and herbivorous forebears, The Dinosaur was a good eater. He *inhaled* food. Actually, he was stingy about practically everything except food, and booze. Then again, this was house Merlot and pizza in Hackensack, not, say, French champagne and pan-roasted Maine lobster with rosemary cream at The Inn at Little Washington. I could only dream about the latter. Maybe for our first anniversary, post-Dinosaurland. But we'd have to get married first. But how could we do that if he hadn't even noticed my preety peenk?

"What's the matter?" I asked.

"That fucking goddamned lying asshole sonofabitch warmonger fool's taking us to war, I just know it."

Oh God, here we go. Dino DETESTED the president. George W. Bush got him *all* hot and bothered—and not in a good way for me. When The Dinosaur started railing against W., who, it appeared, would attack Iraq, U.N. sanctions or no, I knew he would not stop.

"That fucker can TASTE it," The Dinosaur said. "Gotta finish what Poppy started! It's payback time! Gotta demolish Iraq so we can give all those fucking suck-up Cheney Halliburton compadres their corrupt little no-bid contracts to rebuild and—"

"Okay, okay, we're invading Iraq," I said. "Whatever. What about *us*?"

"*Us?* We'll just have to accept it. That bastard's gonna do whatever the fuck he wants to do. Fucking amoral moron. Phony Texas cowboy."

Life was imitating art: There's a scene in my favorite romantic comedy, *Annie Hall,* in which Woody Allen's character, Alvy, is having a political conversation with his first wife, Allison, played by Carol Kane. They're in bed and Alvy starts obsessing over his JFK assassination conspiracy theory. He goes on and on for several minutes until Allison finally breaks in:

ALLISON: Then everybody's in on the conspiracy?

ALVY: Tsch.

ALLISON: The FBI, and the CIA, and J. Edgar Hoover and oil companies and the Pentagon and the men's room attendant at the White House?

ALVY: I-I-I-I would leave out the men's room attendant.

ALLISON: You're using this conspiracy theory as an excuse to avoid sex with me.

ALVY: Oh, my God! (*Then, to the camera*) She's right! Why did I turn off Allison Portchnik? She was—she was beautiful. She was willing. She was real . . . intelligent. (*Sighing*) Is it the old Groucho Marx joke? That—that I-I just don't wanna belong to any club that would have me as a member?

"You don't want dessert, do you?" The Dinosaur asked, wiping his mouth with his sixty-third paper napkin—one for each Cretaceous year of his life, presumably. He added it to the orange grease-stained pile of sixty-two others on his side of the table.

"I'm pretty full," I said, lighting a cigarette. "I'll have coffee, maybe."

"Let's get a coffee at your place," The Dinosaur said, motioning to the waitress for the bill. "I want to catch CNN, see what—"

"What about *us?*" I asked, reaching for his arms. They were beautiful arms. Not too hairy, not too bare. Shapely, strong, masculine. Intelligent arms. The Dinosaur was wearing the black cotton turtleneck I'd bought him for his September birthday. He looks amazing in black, not to mention black shows stains less. He'd pushed the sleeves back to the elbows, revealing the Timex watch I'd given him for his previous birthday. It had a very simple black leather strap and a round face whose black Arabic numerals glowed in the dark against the white background. It was a really good, inexpensive watch. The Dinosaur's pre-*moi* version was a hideous gargoyle of cheesy digitalis that actually cost more than the Timex. It's not how much you pay for something, it's how it looks and how it works—I'm truly democratic in this sense—and, as my girlfriend Eman says, how you carry yourself with it on. Eman ought to know; in September of 2005 that raven-haired sylphette eloped in Cabo in bare feet, a pair of embroidered white lace Felina bikini panties, and a lovely white cocktail-length dress with spaghetti straps. Eman's frock photographed fabulously against her Cabo-caramelized skin and fit her like a dream.

Was it Wang?

La Herrera?

Try Gap.

Just like my little pink raincoat!

"I have to hit the gym early tomorrow morning," The Dinosaur said. "Work off some of this pizza."

"What. About. Us?" I said.

"You want to work out together tomorrow?" he asked, pulling away from my grasp to calculate the tip. "Wow. That's news. You're never up that early. And this ain't yoga, *princesse*, it's not soft-core. You pound it out. You sweat it out. Nothing Zen-meditative about it."

Was this vintage male deflection or could The Dinosaur really be this Dino-dumb?

"Stop talking about the gym and Iraq and CNN!" I cried. " 'Elusive combination of adventure and stability'—Jesus! And my beautiful raincoat, don't you even SEE it? You have no IDEA what it took to get this raincoat. No idea. I worked so hard to get it. It means so much to me. What don't you SEE?"

"I see it," The Dinosaur said, utterly bemused. "I see it."

"IT'S PINK!"

"It is?"

"YES!"

"Oh. Okay. Well, it's pink, then. I'm sorry, I couldn't tell."

"HOW COULD YOU NOT TELL?"

"I'm color-blind."

Peach Panties

Some men reach a certain age and go insane. This is what happened to one Doctor Fruit Cocktail (DFC) when he hit forty-fiveish. By all outward appearances, the guy had a lot going for him: Thriving private practice on a major piece of Washington, D.C., real estate, good health, beautiful skinny wife with perfect skin and really shiny and straight hair, children flourishing in private school, huge mansion he'd built on another major piece of real estate, and biannual two-week-long AmEx-enabled fly-fishing vacations in the Rockies. A rich, rainbow-trouty life.

DFC had been my dermatologist during The Awkward Years, those wretched tweens 'n' teens when no amount of concealer or over-the-counter pimple product or magazine-inspired home-made oatmeal and yogurt remedy could abate my upsetting acne. After a looong-term treatment involving various combos of many topical and oral prescription medications, my chronic facial break-outs eventually calmed down and I stopped seeing DFC.

But. Life is never all smooth for any of us. It may have its moments or drawn-out periods of smoothness. Then things get bumpy again. Expect the bump in the smooth, I always say. I had my front licked; I hadn't thought about my back. Yes, this teen trauma revisited me when I hit my twenties, this time in the form of *bacne*, or back zits. How horrifying is that? They were *huge*. They were *hideous*. They *hurt*. Not only physically, but imagine how badly they impacted my fashion options. This was bad-bad. So I looked up good old DFC and returned to his reassuringly swanky office in the high-rent district of Northwest Washington.

I take any reason to leave the house as a fashion opportunity. You just never know. Some people save their fun clothes for some special occasion that never comes, except in their minds. My special occasion is that I'm alive and I woke up today. So I might as well look cute. It beats the alternative, plus it puts me in a much better mood.

On the bright summer day of my bacne appointment I decided to go with cropped blue jeans and black canvas wedge espadrilles with grosgrain ribbons that wrapped insouciantly around my ankles. On top I wore an ice white cotton short-sleeved Mexican blouse. It was in the indigenous *folklórica* style, with gorgeous, intricately embroidered flowers in bright, happy colors, and little lime green sparrows, mockingbirds, and wrens with real black bead eyes. The blouse was a pullover, with decorative bright pink thread adjustable ties in the cap sleeves, and a deep, scalloped square neckline in front and back. God. Such a fantastic blouse. Like wearing a Frida Kahlo painting but without a freak bus accident resulting in an amputated life, an amputated outlook. My friend Bobbi had bought it and my long silver earrings and my silver bracelet in an open-air *mercado* in Mexico City during a business trip. Bobbi was always going on business trips there. Now technically, ice white and silver are for winters like Bobbi, not for springs like me. I know, you probably think Carole Jackson's *Color Me Beautiful* concept of every person falling into the category of spring, summer, autumn, or winter hues is as dated and old-fashioned as those plug-in makeup mirrors with the tiny frosted ball lights framing them. Well, if they're old-fashioned, then so am I. *I'm just an old-fashioned girl,* as Eartha Kitt sang, *I'll ask for such simple things when my birthday occurs: Two apartment buildings that are labeled "Hers" . . . and "Hers" and an old-fashioned millionaire.*

However, some fashion rules are meant to be broken if you know how. Tip: If you keep your makeup in your correct season's colors, you can pretty much wear anything from the neck down. Hence tomato or coral red—not blue red—lipstick. (I always have to mix, like, at least two different ones, plus gloss, to arrive at lipstick shade perfection.)

Wonderful as my blouse was, this was the first time I could bring myself to wear it for public consumption. Since it was cut low in the back you could see all my bacne and it looked like I had horrendous hives or some ruthless rash. Hormones can suck for a *long* time. So over the blouse I threw on a seasonless cropped black silk knit cardigan with three-quarter sleeves and tiny white pearl buttons. Once I was safely stowed in the examining room, I removed it and my blouse and my bra, as the nurse instructed me to do before handing me a pink paper johnny and leaving.

"Well, look who's here," Dr. Fruit Cocktail said, smiling. He looked ruddy, prosperous, well fed, and incurably self-satisfied. He hugged me and kissed my cheek. "How've you been? You're all grown up. Very nicely, I might add."

"Hello," I said. "And you're exactly the same."

"Not really," he said, his smile fading a bit.

"Oh, you are the same," I said, touching his arm with my hand. "The years have been kind. So? How's the fly-fishing? You and your family must be the only Jews who fly-fish. In *Wyoming.* How less Jewy can you get than that? I saw that picture in the waiting room. You with your son? Holding—what was that—a trout? Rainbow trout?"

"Steelhead, actually."

"Like I would even know this."

"Tough fish to catch," DFC said, opening the back of my johnny to study my bacne.

"Tough nut to crack," I said.

"They spook easy. You have to . . . entice them."

"The zits?"

"The trout. Okay, I see what you're talking about. Yeah . . ."

"Really horrifying, right?" I said. "I'm so embarrassed."

"Nah. I've seen much worse. Bacne's harder to treat than acne, but I think we can fix it."

"Oh good," I said. "Because I was thinking of attacking it with rubbing alcohol. You know, just pour neat alcohol down my back. Straight up."

"I guess it couldn't hurt," he said, smiling. Smiling, but I got the feeling he was making a little fun of me. DFC reached for his prescription pad, and with a shiny, fat black and gold Montblanc fountain pen wrote down one oral and one topical Rx and handed me both.

"These should help," he said, his thick gold wedding band and wavy dirty-blond hair catching the greenish overhead fluorescent light. Maybe he was a Carole Jackson summer. No, more likely a spring. The starched, snow-white lab coat made his reddish facial skin look seriously distressed. DFC's fingers, like the rest of him, were big and chubby but tanned, consequently more attractive than they'd be otherwise. Something about tanned skin, even if you're an orca, always looks okay. Like, acceptable. It just looks more toned and potent than pasty white skin with veins. All those fancy fly-fishing holidays, you know? Me, I get in a car and wade through a Maryland mall to get Clarins self-tan.

We had kind of different lifestyles. He lived in a suburban castle. I lived in a divey studio walk-up that smelled like roasting lamb shanks and had cockroaches in the kitchen, courtesy of the Middle Eastern deli below. My "view" overlooked the parking lot behind the building where the Dumpsters were. Therefore the bugs, I think. As Bob Dylan sang, *What's a sweetheart like you doing in a dump like this?* Good question, Bobby.

"Wouldn't it be nice if all our problems could be solved with an Rx?" DFC asked. He looked a little droopy. But hey, rich people's problems. What can you do? DFC didn't have any problems. None that I could see.

"Well," I said, "mine can."

"I'm about to get separated," he blurted.

"I'm sorry," I replied. Why was he telling me this? *Every*body tells me their personal problems. Why? Troubled married man magnet, *c'est moi*. Why am I here? On Earth, I mean. Am I supposed to solve all the world's woes *in addition* to the *Color Me Beautiful* ones? Really. All I want is no more bacne so I can wear my Frida Kahlo blouse in peace. Jesus! Is that asking for too much?

"My son? The one in the picture? I may only get to see him and his sister on weekends from now on," DFC said. "I keep buying them all this crap because I feel so guilty. Then my wife yells at me. 'It's gonna fuck them up. What're you trying to do, buy them so they'll love you more since you're never here? They already think you care more about your precious patients than us.' "

"Uh-huh, and what do you think is the best thing for chapped lips?" I asked, desperate to change the heavy subject and lighten things up. "I've heard plain Vaseline but do you think something a bit more medicated or, like, mentholated—"

DFC took a step back, assessing the situation. He sighed. It was loud and portly.

"What's the matter?" I said.

"You don't want to hear this."

"What am I hearing?"

"Why don't you get dressed and meet me back in my office?"

It's hard feeling too sorry for rich people. I once told my beloved rabbi, Bruce Kahn, that I wished I was the wife of a guy like DFC. (Bruce is really normal, for a rabbi. That's why I talk to him.) He goes, "Why'd you ever want to be that?"

I go, "Because those wives have charmed lives. They're not as buffeted as the rest of us. They're, like, protected and pampered. I've seen them. They shop at Neiman Marcus at two o'clock in the afternoon. And they look all ironed. Smooth. It must be so great to be that."

"Dear, they shop at Neiman Marcus at two o'clock in the afternoon because they don't have anything else to do and no place better to go," Rabbi Bruce said. "Don't fall for façades. Nobody has everything. Nobody. You have no idea what's going on at home behind closed doors. A lot of rich people are really empty and unhappy."

"I hope so but they don't look empty and unhappy to me," I said. "They look like they just had a lot of oxygen facials. If you have to suffer, it's better their way. I'd like to shop at Neiman Marcus at two o'clock in the afternoon with my husband's money."

"You might like it at first," Bruce said. "But if that was all there ever was? No, that's a fantasy. You're a writer. You have real work to do. You have things to offer the world that those women don't."

Maybe. But those women look a lot better doing nothing than I do. It's hard enough being a woman, it's *expensive*. Poverty is not an advantage to beauty. You have to be a hundred times more resourceful and creative when you're poor.

"I'm so glad you came in," DFC said, leaning forward in his adjustable leather and titanium ergonomic chair. It was a really good chair. I aspired to that chair. I'd seen a chair just like it in Theodore's window. (Theodore's is a very nice furniture store in upper Georgetown.) The rest of DFC's office was like that chair: Masculine, stately, dark, substantial, well designed. Diplomas hung in gold frames on the walls. There were framed photographs of DFC and his family everywhere. Every person in every picture was smiling and robust, like they had really good health insurance they never needed to use. There was one photo of a gorgeous brunette I took to be the Neiman Marcus wife. She looked like Kristin Holby, the Ralph Lauren model who played Penelope Witherspoon in *Trading Places*. In another photo smiled a skinny, tanned young girl with streaked blonde hair, wearing nearly invisible braces, proudly displaying a freshly caught fish on a line. Definite piscatorial motif going on around here.

"My daughter," DFC said, looking at me looking at the picture. His big elbows made twin depressions in an enormous leather desk pad covering the massive cherry desk. I leaned back into my leather armchair. It was tufted.

"She's cute," I said.

"She is," he said. "So are you."

Uh-oh. Here we go.

"Thank you," I said, shifting on the upholstered tufts. "I'll be a lot cuter once I'm bacne-less. Thanks for the scrips."

I got up. The wedge espadrilles gave me some height and as DFC walked around the desk and faced me, half sitting and half leaning on the edge of his desk, we were almost eye to eye. He smiled flirtatiously. It was inappropriate and sort of predatory, but I wasn't afraid. After all, I'd known the guy since I was, like, fifteen—almost a decade. He'd cleared up my acne, belonged to a wealthy synagogue, and put the kiddies in private school—how scary could he be? Besides, he's a *doctor*. First, do no harm. And just how naughty could anyone be with his Neiman Marcus wife's, young children's, and rainbow and steelhead trouts' eyes watching from every direction?

"Pretty blouse," DFC said. "Are those real beads?"

"The birds' eyes? Yeah."

"It's nice. You're nice. Hey, have dinner with me."

"What kind of dinner?"

"Um, the kind you eat after lunch," he said, laughing. "I could stop by your place sometime if you invite me. Pick up some—you like Thai?"

"Eat in?" I said. "You want to eat in?" Dammit. Did I just actually say that? Could we possibly be a tad more Freudian?

"Well," DFC said, "we should probably be discreet. Eat in. I'm still living at home and a lot of people in town know us both. Your place is kind of out of the way but someone might recognize my car out there."

"I'm not that 'out there,' " I said, feeling miffed. I mean, okay, Takoma Park, Maryland, isn't exactly Potomac, Maryland, but still. I'm a struggling, self-supporting, freelance writer trying to make a name for myself, and let me tell you honey, it's one slow bitch of a career pick. *You*, on the other hand, are in *Washingtonian* magazine's Top Doctors issue every time it runs. Not quite the same thing.

Nevertheless, DFC's offer intrigued me. These things always do when you're in your twenties and broke and between boyfriends. Plus it conjured a frisson of naughtiness, which of course made me really want to do it. I'd never been with a married man before but, being basically clueless, I was sure I could handle it. As for our significant age difference, well, that was neither here nor there; I was in a geriatric phase.

I confided my impure intentions to my (then) long-distance gay friend Billy. He said, "You go, you HOO-wuh!" Which is how people in New Jersey, particularly of the Sopranos' ilk, pronounce the word *slut*. It's a regional compliment.

Okay, so maybe I'm a HOO-wuh, but what is this HOO-wuh gonna WAY-yuh? Belying gay stereotypes, Billy doesn't know or care much about clothes. He calls himself "a heterosexual with emotional problems."

So I called my local girlfriend Bobbi to discuss. Bobbi's a well-dressed, fairly dissolute immigration attorney who drives a Jaguar, smokes Marlboro Lights, and loves me.

GIGI: Okay, my little Bobster. It can't be too obvious. Nothing overtly cha-cha.

BOBBI: Exactly. So, like, jeans and a cute top. *(To her assistant's assistant, Marta)* I'm leaving for the airport ten minutes ago, so you need to have that file ready for me, like, yesterday. Thank you, Marta. *(Back to me)* Sorry. I'm in summer intern hell. Okay, so I think we go with a tank or a camisole or something. It's summer. Very casual. You're tiny so it'll be cute. You're lucky 'cause God knows your A.C. sucks. I always sweat like a pig at your place.

G: Dear, *I* always sweat like a pig at my place.

B: I'm buying you a new unit for your birthday. It'll be like an Arctic hurricane.

G: My birthday's in December, so you're probably right.

B: Okay, so where were we? Sorry, I don't have much time. Let's stay on message. I gotta make a plane to Mexico City. I have to go to the American consulate down there because the judge here lost the papers and—anyway. It's a whole big mess. *(To Marta)* MARTA! Señor Hernández's file? Did you load it in my laptop?

G: My outfit. My dermatological outfit. Please focus on message.

B: Right, okay. So a top and jeans, and a nice bra and underwear. And just do earrings. No other jewelry.

G: Okay. I can totally do that. The Ann Taylor acorn drop pearl earrings with the silver hooks? Elegant yet kicky. And no shoes, right?

B: Definitely no shoes. You're at home, you're relaxing. Just the red toenails, as usual, and keep the makeup soft.

G: BOC. [But Of Course.] Soft makeup that will take two hard hours to put on. Oh, and Midnight Red or the corally red one? The Chanel polish for the toes.

B: Maybe Midnight, 'cause it's more classic.

G: I agree. And ambiance-wise?

B: Light a few candles and play something classic. James Taylor or Steely Dan or something—he's old, right? Just don't make it too, like, "Oh, the seduction parlor!"

G: That'd be tough considering *las cucarachas* in the kitchen and roasting lamb shanks for aromatherapy, dear.

B: I know. But you know what I mean.

G: Yeah. So black bra and panties? Isn't that what they all love?

I have this really cute set from the Gap. It's that stretchy nylony opaque 'tude? Underwire with thin straps and low-rise boy shorts? Not too wenchy. It's not, like, default setting to adultery scenario: "Frederick's of Hollywood! Can't think of anything else! Let's resort to clichés! Give me your thongs, your G-strings, your sheer lace bustiers with the nipples cut out!"

B: "Yearning to breathe free."

G: Thank you, Emma Lazarus. Poetess of the immigrant nipples.

B: And instead of standing there in a robe, holding a torch and a tablet, you can greet him at the door in jeans and a camisole, holding a cigarette and a cocktail. Now *that's* poetry.

G: It's all gonna come off in two seconds anyway. The makeup and the clothes.

B: We hope. *(To Marta)* MARTA! *(To me)* Fuck me, now I'm late.

G: However, XY chromosomes are visual. That first impact, it has to be there.

B: It'll be there. He's an XY. You're the statute of libertines. Gotta go! Good luck!

G: This is really bad, right? That I'm doing this with him? Should I not do this?

B: Oh please. One of my partners is fucking her client. Are you kidding? *(To Marta)* MAR-TA! *(To me)* I heard you light that cigarette, missy.

G: *(Exhaling)* But it's a short one, I promise. Parliament Light.

B: Just please tell me you're not gonna cook for him in that steam bath you call your apartment in summer.

G: No, he's bringing stuff.

B: Hope he packed some condoms in the picnic basket. BYE! *(To
Marta)* MAR-TA!

While Bobbi spent three days frantically hunting down con-
suls and Señor Hernández's lost legal papers in Mexico City, I
spent three days frantically cleaning, dusting, and vacuuming my
apartment; going to my blonde German aesthetician Ute for eyelash dying, lip and bikini waxing (hey, I might've been poor but
some things are too hairy not to splurge on, which means they're
not really splurges); doing my pedicure (even with decent A.C.,
my skin is abnormally warm, so it takes two full days for my polish to dry; hence, I do it myself); shaving and self-tanning the
legs; ironing the jeans and cotton-knit coral pink camisole (I iron
everything except my cat Lilly); running to Kmart for white
tealights, those wonderful, weenie little one-and-a-half inch
round candles that come in tin silver cups ($2.99 for one hun-
dred—can't beat that price with a wick), and glass tealight hold-
ers ($1.50 for four; excellent price too, especially since I already
owned a bunch and, as with shoes, the more the merrier); doing
my yoga; buying three blue hydrangeas at Johnson's, my favorite
D.C. florist; and setting the tiny round table in the kitchen with
said hydrangeas in a tall green glass vase on the far end.

Spontaneity takes forever to plan.

On the appointed evening, I was in the bathroom dabbing on
last-minute Clinique lip gloss, smoking a Parliament Light 100,
drinking white wine, and listening to an old Steely Dan CD I've
always liked whose title, *Countdown to Ecstasy*, seemed fitting. I
walked into the main room and peered out the window. A huge
boat of a Mercedes was pulling into the parking lot. Was that
DFC's? It had to be. Only a certain kind of Jew would spend
$50,000 on a German car. I have relatives like that.

He parked as far away as possible from the rusting Dumpsters, got out—he was in shirtsleeves, jeans, and moccasins; must've changed clothes after work—and opened the hatch. He extracted what appeared to be three weeks' worth of groceries in overstuffed white plastic bags. Was he planning on moving in? Did he feel sorry for my impoverished straits and this was his tax-deductible charitable donation? Was he suggesting I was too thin? (No. I may be puny but I've never been too thin, and certainly never too rich.) Was he showing off while slumming? Was he so indecisive in the same way he was "about" to separate from his wife as he was "about" to choose one or two dishes, but he couldn't choose so instead he ordered everything on the freakin' menu? As in, "I want the perfect Neiman Marcus wife *and* the Dumpster girlfriend with the bacne, AND, I want the whole fried fish *and* the whole roasted suckling piglet"? Would that make him the Gordon Gekko of the D.C. doctor set? *Greed is good.* Maybe I had my own, if latent, avarice. How about that? Or was DFC just feeling kind of lost and at loose ends like I was and he was trying hard to be nice and make a good first impression? There's a concept.

I finished my wine with one gulp and pressed my ear to the door. I could hear him arduously climbing the stairs and *grunting*, for God's sake. Didn't he say he regularly played handball "at the club" with a reconstructive plastic surgeon, an otolaryngologist, and an infertility gynecologist? Should I go out there and help him? Would that be too loser-eager? Oh God, what if he had a heart attack and died in the dirty stairwell? All that good food would be wasted! How tragic! Then the *Washington Post* would write about it in the Metro section, and Neiman Marcus would go berserk and I'd be implicated and—

DFC breathlessly knocked on the door.

I opened it and said, "Hello."

"Hi," he said, catching his breath. He was almost sweating

through the embroidered baby blue polo player on horseback on his pink linen shirt. "I made it."

"Come on in," I said. "Wow, looks like you got a lot of stuff there."

"Where's the kitchen?"

"That way," I said, pointing.

DFC piled everything on the counter and leaned against it with his hands.

"Nice to see you," he said, turning to me. "You look nice."

"Did you just run a marathon? Would you like an ice water?"

"That sounds great."

"Sorry about the steps," I said, twisting an ice tray. "I should've warned you."

"Well, when you said, 'I'm on the first floor,' I thought you meant ground floor."

I handed him a tall glass of ice water and watched him guzzle it. I poured him another glass. He drained it and held it to his forehead. It was a virile gesture.

"Thanks," he said. "So. This is your place."

"This is my hot little place, yes. Sorry, the A.C.'s on full blast but—"

"It's what I imagine a writer would live in," he said, surveying the main room and slipping off his black shoes. They were Tod's. Figures. "You know, arty. I like that sloped ceiling. It's like being in a charming little Parisian garret. Cozy."

"Thanks," I said, discreetly checking the floor for *cucarachas.* What if some crawled inside those Italian leather moccasins? I wish I had those moccasins.

"So shall we have a glass of wine?" he said, walking back into the kitchen and rummaging through the twenty-seven bags. "I brought beer, too. Oh, here it is. I'll stick this in the freezer. I'm still kind of hot."

"Why did you bring so much?"

"I don't know," he said, shrugging. "Just felt like it. Wasn't sure what you liked. The only wine I like with Thai food is a California gewürztraminer or maybe something from Alsace."

"I usually just have iced tea, myself. But wine is good."

"I brought a Kabinett and an Auslese. Rieslings."

"Okay. I have no idea what you just said, but okay."

"Spicy wines. The spices in Thai food tend to overwhelm almost all other wines. Lots of people chuck the wine and just do beer. I wasn't sure which you preferred."

"You could've asked," I said, handing him a corkscrew and two wine glasses. DFC's cork-removal technique was suave.

"I haven't done this a lot," he said, handing me a glass.

"What, uncork a wine bottle?"

"No."

"Oh. So are there appetizers here? Let's pull out some and sit down in the 'living room.' I use the term loosely."

There were grilled chicken wings in toasted coconut, and steamed dumplings and beef satays with peanut dipping sauce. This introductory selection did not include the tomato and shrimp salad with lemongrass and anise basil, the hot and sour soup with shrimp toast, the cellophane noodles *and* sticky rice with vegetables, the pla rad phrik (a deep-fried whole fish—note the piscatorial motif—with an intense chili sauce), or the pink grapefruit sorbet with mint. There were a couple other bags in the heap; DFC said those were a surprise for later.

Fortunately, I have a ridiculously vast number of good serving dishes and wares: Dinner-, flat-, and drink. For an unmarried youngster with no money, I had a "very well appointed" kitchen, as my British friend Marcy once remarked most Britishly. "Now all you need's a husband!" she said.

Well, now I had one. He was sitting right next to me on my

Conran love seat and tearing into truly riveting beef satays. Everything was sensational. DFC was spoiling me silly. A person could get used to this, especially poor me. After a second or maybe a third glass of Kabinett and three dazzling dumplings in peanut sauce, we kissed. The old guy was good. Very good. Thai-good. He got up, extending his hands for mine, pulled me to my feet, and wordlessly tossed me in bed. He devoured me like a grizzly bear would a cutthroat trout, bones and all.

Afterward, I put on my hand-embroidered silk kimono—another indulgent present from Bobbi's Mexican travels—and served up the sorbet first, to cleanse our palates, so to speak, and then the rest of our fabulous food. DFC rummaged through the other bags and withdrew a packet of rolling papers and a quart-sized Ziploc Baggie packed with Hawaiian weed. Whoa. *Weed? Him? Really?* Really. My dermatologist had unsuspected depths—or shallows—and the bucks to afford the cream of the islands' crop. We shared a fat joint and truly, it was the most amazing weed I've ever smoked, and I'm not even a fan. DFC placed a heavy plastic grocery bag on my lap, which contained several tons of skin-care product samples, including countless tubes of pricey Differin 0.1% cream for my bacne. I could be cleansed, exfoliated, toned, clarified, comforted, purified, peeled, plumped, moisturized, masqued, hydrated, brightened, softened, lightened, evened, calmed, renewed, detoxified, refined, unlined, de-wrinkled, firmed, anti-aged, nutrified, smoothed, soothed, revitalized, replenished, restored, uncreased, tightened, radiated, relaxed, pore-minimized, sun screened, and bacne-free for the remainder of the millennium! This was better than sex. With anyone.

We tucked into the rest of the Thai bacchanalia and then fell back into bed. What had he told Neiman Marcus as to where he was? I wondered. I was about to inquire when DFC decided to go dowwwntowwwn on me—and *quite* expertly, too. So much so that I forgot I even had a question.

After a while, DFC came up for air.

"You're delicious," he said. "You taste like fruit cocktail!"

"Would that be canned or fresh?" I asked. "Because there's a significant difference." Even in my semen-infused, drug-addled haze, I knew that much.

"I don't know!" DFC said, thereby sealing his fate, not to mention new and permanent sobriquet. If you don't know the *obvious* right answer—FRESH, which would actually make it fruit SALAD, not CANNED FRUIT COCKTAIL—then *you're* a FRUIT COCKTAIL.

Dr. Cocktail showered. I languored across my bed like a kitty cat, stretching my arms and legs all the way out like I was making a snow angel. Even though Peggy Lee once sang "Don't Smoke in Bed," I lit a cigarette and knocked back an entire liter of Evian. Mmm. Oh my God, I felt *goood*. I looked at my white reflection in the black window. Even with my makeup worn off, I looked pretty. Happy, even. You don't have to be married to them to get this glow. (In fact, it probably helps if you're *not* married to them.) Maybe I was a French mistress in another life, to a Sun King who liked Thai trout!

DFC came over to me and sat on the edge of the bed. His hair was wet and his big red face was serene. He handed me a business envelope.

"Take this," he said. "I want you to have it."

"What is it?" I asked, smiling.

"It's a gift and a favor. Open it."

The envelope contained five $100 bills. I had no *idea* Ben Franklin could be so desirable.

"I know you're struggling," he said. "This is . . . this is an endowment."

"Wow," I said, grabbing him and kissing him on the mouth. "I don't know what to say." But of course I did: "Thank you."

"You're welcome. I have a favor to ask. I want you to go out and buy yourself some serious lingerie. I mean, top-of-the-line, luxurious silk and lace. Something imported. A bra and panties and a garter belt—"

"A garter belt? You're serious?"

"And stockings. Actual, real stockings, not panty hose."

"Do they still even *make* those?"

"I want to see you in those things and ravage you. Oh, and one more thing: It has to be peach. The lingerie has to be peach."

"You don't like black?" I asked, feeling suddenly clichéd. "What's wrong with black?"

"Nothing," he said. "But black's too severe on you. You're fair, you're . . . soft. You need something tender. Gentle. Peach."

"Is this because I taste like fruit cocktail? We've got a fruit theme going on? 'Do I dare to eat a peach?' "

"See you next week, Peach. Thursday okay?"

"It's so okay," I said. "It is *so* okay."

"*Oh my* God!" Bobbi shrieked. She was back from Mexico City and we were sitting outside Sutton Place Gourmet in Bethesda, Maryland, so we could dissect DFC while chain-smoking. "He wants you to spend five hundred fucking dollars on fucking UNDERWEAR?!?"

"Excuse me, Miss Jag Queen," I said, sipping my toffee nut cappuccino. "He never specified the ENTIRE five hundred dollars was allocated for it. Maybe he has a fetish."

"Go to Target and pocket the leftover 490 bucks," Bobbi said, wiping off her frozen hot chocolate with caramel and whipped cream mustache. "He'll never know the difference."

"Oh, this one would *know* the difference. Trust me. The guy wears *Tod's*, for God's sake. How gay is that? He's a total brand-name hooker."

" 'Hooker.' That's funny. Doesn't he fish? He's caught you, all right. Girl on a line. The Ancient Mariner reeled you in."

"Ancient Mariner's cunnilingus technique's *breathtaking*," I said, dragging on my Parliament. "You cannot disregard. Not to mention enough Thai leftovers for a week. Not to mention massive product samples. *And* Hawaiian pot, which, unlike the samples, I will gladly share with you."

"Guess it's time to hit Neiman Marcus," Bobbi said, finishing her drink and our dissection with an emphatic, caffeinated slurp.

The following day, I was shopping at Neiman Marcus, aka Needless Markup, at two o'clock in the afternoon with somebody else's husband's money. In my fantasy, it was great. To my surprise, however, I didn't feel charmed, protected, pampered, ironed, smoothed, or facially oxygenated. I was actually kind of anxious and sweaty because I couldn't find a single peach piece in their entire intimates department. Not one. What a bust! My five fat Franklins were burning a hole in my wallet—and *nada*. What's up with the white-black-nude-only obsession? There were some great fishnet stockings in black and white—I love fishnets—but nothing in peach. And peach was my mission and I had chosen to accept it. Meaning I was determined. Meaning I wanted to keep my *lagniappes* coming. I could get into this gift-with-purchase attitude. Why not? Hence, this search was no chore; it was a barter, it was fun escapism.

I hit Nordstrom, where I found nothing in peach either, although I did get a nice pair of Hue elasticized lace-top, thigh-high stockings in "natural" (close enough) for a mere $7. Wearable with or *sans* garter belt. Was DFC really gonna be able to tell they were natural and not peach? Was he really that gay?

Thank God for Saks. Saks saved my ass and made me feel truly oxygenated. Their lingerie department rocked, and for a paltry $135, I was all set in poshest peachiness: A beautiful Le Mystère lace tulle underwire bra and matching lace panty with tiny satin bows on the respective fronts. I loved Oroblu's Italian satin and lace garter belt, but they only had it in white, so I stayed in Gaulville and bought the peach Le Mystère one. In the full-length mirrors, I thought I looked romantic in a Moulin Rougey way. Toulouse-Lautrec would've loved my cancan. Would Dr. Cocktail? *Bien sûr*! And I had to admit it: Peach did look better against my skin than black.

"Oh, cash," the saleslady said as she rang me up. "You don't see this too much anymore."

"Ben is my friend," I said.

"These are beautiful items," she said, wrapping them individually in tissue paper. "Lavish. Are they for your trousseau?"

"No, just for my beau. I'm actually not that into lingerie."

"He's a lucky guy," she said.

"Oh yeah, he got lucky," I said. A lucky guy, like the melancholy Rickie Lee Jones song: *He doesn't worry about me when I'm gone, he goes to sleep at night, he don't turn off the light and wonder how to find me or if I'm alone . . .*

Predictably, DFC *loved* my French ensemble and ravaged me, as promised, on that night and on many subsequent Thursday nights. He always brought the great Thai food, the amazing Hawaiian weed, the hundreds of skin-care product samples, and a few Ben Franklins for me to buy replacement stockings for all the pairs he wrecked. (And no, he never said anything about them not being precisely peach.) After several months of DFC's treatment, my bacne cleared up, too, thanks to all that complimentary Differin. DFC and I were in a comfortable routine. The only change I detected in Dr. Fruit Cocktail was that lately he had stopped complaining about Mrs. Fruit Cocktail. Before, he only mentioned her to criticize her. Now, he never mentioned her at all. Maybe DFC was falling for me. Could he be?

He phoned one Wednesday, sounding atypically somber. He wanted to meet in his office the next day at six.

"In the *morning*?" I asked.

"No," he said. "At the end of the day, after everyone's gone home."

"Mmm! In your *office*? I love it. How *wrong*." Hey, whatever floated his fly-fishing boat. I heard him sigh his familiarly loud, portly sigh and we said good-bye. I attributed DFC's tone to his being surrounded by patients or staff.

The next day I donned my fancy scanties (freshly hand-laundered with Woolite in the sink and drip-dried, as always) and black stilettos, and I wrapped myself like a gift in a belted black trench raincoat, knotting the strap at my waist. I wore nothing else (unless you count the faux diamond studs, black sunglasses, and red lipstick). I cautiously crossed Connecticut Avenue at rush hour, praying not to get hit by an oncoming car. That would be gauche to explain in the ER, wouldn't it? *Yeah, um, I was on my way to rendezvous with my married dermatologist lover and . . .*

DFC kissed not my lips but my cheek—clue number one something was off—and ushered me into an examining room, the same one I'd been in all those appointments ago. I sat on the paper-covered table and crossed my legs. My heart was racing.

"Thanks for coming in," he said, closing the door.

"I almost got run over down there," I said.

"Why are you wearing a coat? It's, like, eighty-five degrees outside."

"You're right," I said, slipping off the table and unfastening my belt. "I shouldn't be wearing a coat." I let it drop and stood there in all my Mystère.

"Jesus," he said, rushing to close the Venetian miniblinds. "What the hell?"

Clue number two. I sat back on the table, suddenly feeling ludicrous among the sterilized jars and drawers and cabinets of medical supplies and DFC's stiff white monogrammed lab coat. I'd never noticed the monogram before. It was like living out one of those dreams where you're in a crowd, and it dawns on you that you're the only one who's naked.

"Problem?" I asked.

DFC sighed. Third clue's the charm.

"We have to stop," he said. "I'm sorry. I can't do this anymore."

"Why?"

"I have cancer."

"You *what*? Where?"

"I need to live at home with my wife and my children and get treatment and concentrate on beating it."

"Oh my God," I said, getting up to hug him. "I'm so sorry! That's awful! Is there anything I can do?"

He picked up my raincoat and helped me put it on.

"I guess prayer couldn't hurt," he said, opening the door. He said it kind of smiling, like the way he'd made fun of me for wanting to pour rubbing alcohol down my bacne'd back.

"Wait," I said. "So . . . that's it?"

"That's it," DFC said with a nod.

"Just like that? I mean—"

He hugged me and I wished him luck. I thanked him for everything he gave me. I told him he was a good dermatologist and we both started crying and I left.

"*Fuck!*" *Bobbi* said. "Fuck!"

"I know," I said. We were in her living room the following afternoon, drinking TaB and Diet Pepsi and smoking cigarettes and a leftover half of a Hawaiian joint. I'd sounded so upset on the phone that Bobbi had taken off a half day just for me. That is true friendship.

"How do you know he's telling the truth?" she asked. "Maybe he reconciled with Neiman Marcus and he's too gutless to tell you."

"Like he'd lie about cancer, hello."

"Well, we know he's a liar. He was sleeping with you while wifey was at home. He's a player. Plus, he fucked his patient. Those people will do anything, they'll say anything to get rid of you once they're tired of you or the guilt kicks in or whatever the hell it is."

"I bought a sixty-dollar bra and thirty-dollar panties for this person," I murmured, "and I'm still alone. What was I thinking? I knew I could never be Mrs. Neiman Marcus."

"Come on," Bobbi said, putting out her Marlboro Light. "Get up. We're going for a drive. We're being proactive."

"What? Where are we going?"

"We need closure. It'll clear your head. That thing in his office, that wasn't closure. That was, well, I don't know what it was. If this is the end of the line, if this means no more gravy train, then fine, but let's see for ourselves. You need a reality check."

We got in Bobbi's Jag, convertible top down, and sped to the site of the huge mansion DFC had built. His street was a quiet, leafy cul-de-sac lined with other perfect mansions with other luxury SUVs in the driveways.

"Keep your head down," Bobbi said, parking a few mansions away from DFC's gilded front door. "I'm gonna pretend I'm lost and ask for directions."

"WHAT ARE WE DOING HERE?" I screamed.

"Investigating reality!" Bobbi said. "Now keep your head down and be quiet!"

"Stop bossing me! You're freaking me OUT!"

Bobbi walked to the house and rang the bell. I saw a woman who looked exactly like the one in the photograph in DFC's private office answer the door. She was tall, slender, and tan, with glossy chocolate brown hair down to her shoulders. She was wearing a sleeveless pale yellow pullover and floral knee-length shorts. She let Bobbi in and closed the door. I dabbed on some Clinique lip gloss, lit a Parliament, and waited. Fifteen minutes later, Bobbi emerged, and we drove off.

"The wife is a knockout," Bobbi said, lighting a Marlboro Light. "And really nice, I hate to tell you. See? Men—they're all dogs."

"And?"

"I asked to use the bathroom and it's frescoed."

"What?"

"The walls, they're hand-painted frescoes of the Mediterranean."

"Must be nice."

"And the two kids had some friends over. They were out by the pool."

"Of course they were."

"He's a dick, okay? That—that *place* is not for you."

"It could've been," I said.

"No, it could not," Bobbi said. "You'd get bored very quickly."

"Now you sound like Rabbi Bruce."

"Fruit Cocktail made the right decision for both of you."

"Say it again. Maybe it'll penetrate."

"He made the right decision. That was just a fling, you know? Fuck him. You got some good shit out of it."

I went home and put away my peach silk and lace lingerie for the last time, tucking it carefully in the bottom of my underwear drawer. That way it's a memento, a trophy, and not a constant reminder of what I lost, and I did lose something with Dr. Fruit Cocktail besides my bacne and some extra income. Unlike my little pink raincoat, I've never worn those things again. That would be too weird. But it's still really pretty lingerie.

Red Ballet Slippers

Part One: MIAMI

When I was little I took ballet, and ever since I've loved ballerina slippers and shoes that remind me of my first baby pink pair. Flats may not literally lift me up but they do turn me on. This is unfortunate, as I'm only 5'2". I've always liked going counterintuitive. While I fully grasp their elevated sex appeal and eternal elegance, high heels for me are usually high hells. I have medium-wide, easily abused feet and there's nothing less sexy than your feet killing you. If I could go barefoot all the time, I would. A pair of painless heels, a pair of heels that, God forbid, feel good *and* look good—that's my Holy Grail. Maybe they're a myth, like unicorns. Stilettos give me multicorns. I think for a woman to be happy she has to be comfortable. Podiatrically, especially.

So over the years I've amassed a staggering assortment of fabulous flats and quasi-flats. They keep me down-to-earth and stylish and release me to run and have fun. My girlfriend Eman, also a flats fan, says we love them because of "the C's: comfortable, casual, cute, carefree, classic, cavorty, convenient, charming, chic, cheerful, and commonsensical." To her C list I would add the F's: friendly, French, and un*f*orced. Okay, that last one is not strictly an F, but you get what I'm saying. A woman in ballet flats is confident and kicky, sweet and free. Her sex appeal does not depend on elevated inch count. It's just *there*.

I was hoping my blind date, whom I dubbed The Snob, would pick up on this flat-footed idea and have the sensibility to appreciate it, and therefore me. *Of course a girl in ballet flats is the best kind of girl to love!* My old childhood friend, Graciela, had set us up. Actually, it was her husband, Jorge. The two had known each other for years, having grown up together in Argentina. *¡Muy exótico!* Though I'm Cuban-born and Spanish is my first language, I never dated Hispanic men. Where I was raised, in Washington, D.C., there weren't any around except for my dad. So the thought of dating a real live Latin lovah was a novelty for me.

"That's why it'll be fun!" Graciela, who's also a native Cubana, told me over the phone as we hatched our intracontinental scheme. She and Jorge and their two little kids lived in Miami. "It'll be like coming home." Considering I'd spent my life trying to get *away* from home, I wasn't sure that was such a hot idea. "I can just see it," Graciela said. "You two will have the cutest babies!"

In my mind's eye Graciela transmogrified into Reizl Bozyk, the cartoony actress who played Bubbie Kantor in the 1988 movie *Crossing Delancey*, and she launched into *The Fiddler on the Roof* lyrics: *Chava, I've found him, will you be a lucky bride! . . . You've heard he has a temper, he'll beat you every night. But only when he's sober, so you're all right.*

It was terrifying. Almost as terrifying as the fact that I agreed to fly to Miami *in August*—AUGUST, AUGUST IN MIAMI—for one blind date. More than a thousand miles. For one blind date. At least the airfare was cheap. Sure, sure, I was going to visit Graciela and Jorge, whom I hadn't seen in a while. And yeah, yeah, I was glad to have a little getaway. But we all knew that the main attraction was The Snob. You can find people to date in our nation's capital, sort of, but The Snob sounded like more than just someone to date. I'm not alone in my D.C. dating drought opinion. Oprah once said that the reason she left her job in Baltimore (which is only thirty-eight miles from downtown Washington) and took what would be the fateful one in Chicago was because her girlfriends told her she'd be more likely to find an eligible bachelor in the Windy City. And look at Oprah now. I mean, okay, she's not married, but that's by choice. And she's richer than God.

On the plane ride down to Miami it occurred to me that perhaps Oprah's girlfriends were referring to eligible *black* bachelors. Which is odd, considering I've dated plenty of 'em. Never had a dearth of black guys. They love me. I think they enjoy my ebullient personality and fashion sense and passionate nature. Also, kissing a woman wearing red lipstick doesn't throw them, not in the least. Some men are just more enlightened, I guess.

I'd had a recent conversation about all this with my best friend and *Washington Post* colleague Goddessina (short for mini-Goddess or Goddess-in-training). We were baby copy aides together in the Style section. In between answering phones, sorting mail, and writing freelance fashion pieces, we had all these great talks about our little love affairs with clothes, men, and sex. We had this (only semierroneous) theory that the smarter the guy, the dumber—or tamer—the sex.

"Sometimes you just want to get *laid*, you know?" Goddessina said. "No-holds-barred, shake-the-roof, do-me-silly SEX. The kind of sex where you don't care if something rips."

"Yeah, in fact it's hotter if something rips," I said. "Nothing ever really rips, though."

"None of this polite intellectual crap: 'How delicately delicious were those imported scallops? You know, the latest Supreme Court decision is incredibly—' "

"Totally," I said. "But then there's other times when you'd do anything for a polysyllabic conversation, anything beyond 'Hey baby hey baby hey baby' and grunts."

"We need it all in one man," Goddessina said. "*That's* the man for us."

"Does he even exist?"

"Well, *we* do and we're like that. Here, I've got the ideal: The cool of black guys, the warmth of Latinos, and the brains and bucks of Jews."

"Brilliant!" I said. "A multiethnic approach. *Inclusive* love."

"*Mais oui,*" Goddessina said.

"A Jew would never rip your clothes—off or on. But he'd never complain that it takes us an hour to do our makeup," I added. "Blacks and Latinos, too, they wouldn't care. It'd be, like, standard for them."

"Exactly. And emerald-cutting them will get us *our* emerald cuts."

Goddessina and I had developed our own girl code, a feminine friendship shorthand. "Fellatio" had been euphemized into "emerald cut," as we believed the former would lead us to the latter. As in engagement diamond ring style. Hey, it worked for Goddessina. As it never has for me, I can only assume that Goddessina, being Goddessina, has a superior technique—though we've never actually compared, *boca a boca.* I probably needed to bone up, as it were; my emerald cut wasn't cutting it with my flame of the time. *My old flame, I can't even think of his name.* What I *can* think of is how a few times he'd actually PASSED when I'd offered emerald cut, something he *never* did when it came to any other of my personal sacrifices. God, how lousy could I *be*? Did I really suck at head?

"You didn't ever bite it, did ya?" Goddessina asked, lighting a Marlboro Light. Ah, the good old days, when you could Actually Smoke Inside.

"No!" I said. "I mean, not on purpose."

"Well, short of biting it, there's no way you can suck at it," Goddessina said, exhaling a coil of lip-smacking smoke. "So to speak."

"So then if it's not that," I said, reaching for my Parliaments, "what is it?"

"Sweetie."

"Yeah?"

"He's an idiot."

I was so relieved! It wasn't my head! Goddessina the Oracle had spoken. In her omnipresent Chanel Star Red lipstick. Therefore, she had to be correct. That rich, deep, saturated, cool, blue-red was so devastating on her, so *right*. With Goddessina's milky white complexion, dark eyes, Italianate profile, and dark curtain-rod-straight hair lopped into an asymmetrical, insouciant bob, she was like a young French Audrey Hepburn from Italy. I could never wear Star Red. I tried it on once in the newsroom and I looked like Bette Davis in *What Ever Happened to Baby Jane?* You think that's bad, listen to this: This majorly bitchy queen who worked in the Food section happened to pass by and remarked, as only a majorly bitchy queen could, "Oh, God. It's Lady Bird Johnson."

That, like my blind date with The Snob, was almost twenty years ago. The good news is, I don't hang on to negative crap. No, I don't. I let it go. I'm evolved that way; it just rolls right off me. But I will give Queen Cruella this: When you're wearing the wrong shade of lipstick, you will look like Lady Bird. Or Big Bird. Or Bette Bird as Baby Bird. But in the *right* lipstick shade, there's no such thing as overkill. You won't remind anyone of any bird, no avian allusions. You'll easily be able to wear tubes and tubes and tubes of it, PLUS gloss—and it will always look pretty. I could *eat* Gladiator, it's so good. That's *my* very best red lipstick, my absolute favorite. It's Chanel's Star Red for gals like me: a warm, corally, tomato red for peachy-toned, fair-skinned redheads with green eyes and freckles. Gladiator. Isn't that the most perfect name you've ever heard of for a lipstick?

Down in Miami, I'd be Gladiating my date on a sultry Saturday night. The Snob wasn't a Miami resident. He was just using the excuse of visiting friends there, like I was, in order to meet me. He'd moved to Memphis, Tennessee, where he'd gotten a research grant at St. Jude Children's Research Hospital. He was a pediatrician specializing in children's cancers. So noble. I never really pictured myself as the Delta housewife type, but you never know where you might be happy. Besides, I'd be a *doctor*'s Delta housewife. That's different. So I kept an open mind and a cosmetic bag full of Gladiator. I really hoped The Snob was Gladiator-worthy; Graciela had made him out to be a golden boy. "*Oro puro*" was how she put it. Pure gold. So with The Snob I could be entering a golden age!

On the other hand, Nothing Gold Can Stay. That's the title of a great Robert Frost poem that begins:

Nature's first green is gold,
Her hardest hue to hold.
Her early leaf's a flower;
But only so an hour.

Was The Snob too good to be true? He had to have a fatal flaw. They all do.

"The guy's supposedly that fabulous and nobody's grabbed him?" I told Graciela.

"We could say the same about you," she replied.

"I've never denied there's something wrong with me," I said. "I *know* there's something wrong with me. That's the difference. I know it."

Graciela said The Snob was tall (well, for a Latino), dark, handsome, spoke English as well as several Romance languages fluently, had impeccable manners, dressed right out of *GQ*, and lived in a beautiful town house. He was thirty years old, never married, no kids.

"See?" Graciela said. "*Oro puro.*"

"Okay," I said. "But who is he as a person? I mean, beyond the *GQ* clothes and everything. Because the *GQ* clothes come off and then you're just with *him*."

"That's what the date's for," Graciela said, "for you to find out who he is. We just wanna marry you off so you can be as happy as we are."

"Exactly," Jorge said. He'd been listening on the extension. "You're too cute and sexy to be uncoupled, *ché.*" Graciela's husband liked saying *ché* as much as he liked the ladies. (*Ché*'s sort of like saying "you" or "uh" or "hey" or any other conversational filler. It's also how Ernesto "Ché" Guevara got his nickname, because he said it so much.)

"Can't wait," I said, trying not to sound too sardonic. Jorge flirted constantly, shamelessly, with me and with every female, and right in front of Graciela, who shrugged it off in a kind of "That's the way men are, but this one always comes home to me" stride.

"He's rich, too, *ché*," Jorge said of The Snob.

"Really," I said. "As filthy as you?"

"My husband's not filthy," Graciela said with a little laugh.

"Oh he's filthy," I said. "Fil-thy."

I know women who say things like "I'm really hard on my shoes." Well, I don't and I'm not. Every pair I own, the flat ones in particular, are too treasured, every one has a story behind it. The story of the red ballet slippers, for example. First, you need to know that I'm a keeper. Which is not the same as a hoarder. Hoarders hoard to hoard. They're not emotionally attached to any object. They just like a lot of *stuff.* And then there's the opposite: Women who treat their belongings as disposable, hence cherishing none. That's the difference between a hoarder, a consumer, and me. I have *relationships* with my shoes. If any shoe of mine, no matter how minor, ever dies, I am bereft. That's why it's a good idea to buy two pairs of whatever shoes you love. A gal needs backup. Sometimes, though, there is no backup available. Just last Christmas, for example, I put on these padded black slippers I bought on sale at least a year ago from *www.keds.com.* Keds don't just sell sneakers, although I love those, too, and have many pairs. I even have a canvas pair with a print of tiny shoes all over 'em. Keds also sell leather booties and pad-around-the-house slippers. My slippers were made of a faux suedey-velvety fabric, soft as down pillows, and with hard corrugated rubber soles. I *loved* those slippers and bought the last pair they had in size 7½. The slippers reminded me of driving moccasins, another version of flats I adore and have many pairs of, primarily because I can wear them *sans* socks. I was about to go take out the garbage when I felt something drag between my soles and the floor. I looked down and . . . oh no. My sweet little slippers were becoming unsoled from the front of the toe box. The soles were peeling off in desiccated, cracked black strips. Soleless slippers aren't slippers. They're just . . . nothing. There was no point in having them resoled; they cost, like, $15. I had to face it: My adorable, cheap little black Keds slippers were goners. A person needs closure at a time like this. Closure with music involved. Something appropriately elegiac. For the tender taking off of the slippers I chose Vladimir Horowitz—Jews are so good at heartbreak, it must come naturally—playing Beethoven's Moonlight Sonata. I segued into Arthur Rubinstein for two Chopin nocturnes, one for each slipper as I reluctantly, solemnly, and sorrowfully dropped them into the white plastic garbage pail. Plop. Plop. So sad. I blew my nose and wiped my wet eyes. I wrapped it all up with Mozart's Requiem. As the orchestra swelled to its climax, I turned up the volume, opened my apartment door, walked the three steps to the garbage chute, and said, "Godspeed, little Keds. I'll never forget you." And with a wind tunnel velocity WHOOSH, my funeral Mass was complete.

I scanned my bedroom's wall-to-wall shelves of shoes and shoe boxes with their contents Polaroided and Scotch-taped to the fronts for the perfect going-out-for-a-nice-dinner-in-Miami pair. I'd build my blind date outfit around them. That's always the order I go in. Shoes, then outfit.

- Weathered Dan Post cowboy boots? (Champagne leather shaft with hand-tooled black and camel top-stitching, topaz-colored Teju lizard vamp, classic cowboy heel.) Fabulous but way too bulky to pack. Besides, they'd be too hot in that humidity and too informal for a nice dinner.

- Pigskin slingback huarache flats with matching pigskin bows on top? (I found them in a shoe store in Ocean City, Maryland. Can you believe it? Cool shoes in honky-tonk tacky-tacky-tackaaay DelMarVa?) ADORABLE but probably, like the cowboy boots, too casual.

- Milly pewter metallic T-strap "barely there" Bernardo sandals? Palm Beach and Lilly Pulitzer classic for sure, but too open—what if I chipped my Gladiator red pedicure and didn't have time to fix it?

- Tomato red leather ballet slippers with quilted gold tips? *Ding-ding-ding!* We have a winner, Don Pardo!

Hold it.

I don't own red ballet slippers with quilted gold tips. But I *should*. They'd go so perfectly with my Gladiator red lips. Shit. My flight's tomorrow. No way there's enough time to scout out my fantasy flats. Would somebody please invent the Internet? And while we're at it, cell phones? God, the '80s were primitive.

Okay. Back to scanning my little friends:

- Black lace ballet slippers?

- Kiwi-green suede ballet slippers?

🌸 Spectator ballet slippers in white leather with black lizard vamps?

All cute. And all wrong for what I've got in mind. What else? Ohhh.

🌸 Buffed gold metallic Joan & David ballet slippers with grosgrain ribbon ties?

Mmm! Buffed gold metallic for Miami. This could work! Not as orgasmically as red ones with quilted gold tips (note to self: Buy red flats upon D.C. return *immédiatement*) but, no time. Future pediatrician husband awaits.

For pediatric conquest and my shoes' outfit, I envisioned clean, simple, soft, timeless, and understated. I was in a monochromatic phase—whites and off-whites. Things would change by fall—my sartorial passions are inspired and influenced by diverse stimuli—but summer to me that summer meant old-fashioned *romance*: Blouses and long skirts in bleached cotton and cotton gauze. Lace. Silk. Linen. Chiffon. Eyelet. Full, flowy, floaty, fluttery, feminine things. Vaguely costumey. Ethereal. Unfussy. Nothing too structured, crisp, or "new." And nothing black. I'm not big on black for summer. (Note that tennis players always wear white—it attracts the sun less. Black just draws the heat.) And, though white may look like the absence of color, it is, unlike black, every color combined. And so is off-white, almost. Dressing this way also makes packing much easier and lets me rock out with wonderful accessories like Gladiator red lips and slouchy Helen Kaminski straw hobo bags and pearls everywhere. I've always loved pearls, especially when they're a little messed up and piled up. Nothing precious or earnestly WASPy-preppy.

My whites on white and off-whites on off-white and whites on off-white were seasonally and geographically appropriate, pretty, relaxed, subtle, and airy. A little Annie Hall in Santa Monica, a little *Brideshead Revisited,* a little Meryl Streep in *Out of Africa.* And in *Out of Africa,* Meryl's lover was Robert Redford, when he was still Robert Redford. I wondered if The Snob owned a broken-in brown leather bomber jacket like Redford's in that movie. Of course, Meryl and Robert were doomed—he died—but that was a really good jacket he wore to his plane crash.

Incredibly, The Snob showed up for our date in a broken-in brown leather bomber jacket. In August. In Miami. He was willing to suffer for fashion! *Comment cinématographique!* He could be my Robert Redford destiny! And then he'd die in Kenya's Great Rift Valley and I'd go on stoically alone to write best-selling books! He kissed me hello warmly yet formally, the European way, on both cheeks. His manner was courtly, old school, old world. Maybe this was the Argentinean version of *my* movie, if life is a movie: *Out of South America.* The Snob was attractive and smelled like expensive woods and spices. His caramelly complexion and curly black hair were complemented by his starched white dress shirt. He wore it with gray dress trousers, a brown alligator belt, and brown suede loafers with nickel horse bits on the tops. And no socks. The Snob was a metrosexual two decades before the phrase was even coined.

I half expected to see a curbside coach; instead it was a rented black BMW. The Snob drove us to The Forge for dinner, nudging me out of *Out of Africa* and into *Dynasty.* He'd arranged to have us sit in the ornate restaurant's vaunted wine cellar. Was that Henry Kissinger with the Shah of Iran? *Henry! Where have you been? I haven't seen you since Portofino!* It was over-the-top grand, cheesy, and baroque. Not Versailles but *like* Versailles—crystal chandeliers, mahogany walls, stained-glass panels, paintings and statues of zaftig and naked cherubim and ladies, and thousands upon thousands upon thousands of bottles of wine—on Arthur Godfrey Road in South Beach.

"I'd have settled for Versailles," I said, referring not to the French palace but to the casual Little Havana restaurant named after it. "This is . . . wow. Maximalist."

"I work hard, I like to indulge myself on vacation," he said. "Don't you? We deserve indulgence."

"Ah yes," I said, carefully arranging a heavy white linen napkin the size of Texas on my lap as I imagined Joan Collins would. "The simple pleasures."

"We'll start with the Louis Roederer Brut Premier," he told the waiter. And then, to me, "I'm sorry, you do enjoy French champagne, don't you?"

Sure, dear. I drink it all day. This date was working out so far. Bring on the Brut, baby. I should date doctors more. Journalists are all so poor.

"Tonight's special, *ché*," The Snob said in his distinctly non-Cuban Spanglish. "You're everything Graciela and Jorge said you'd be."

"Oh God, which is what?"

"Cute, *ché*. Funny. Sophisticated. A toast, shall we? Let's toast to the beginning of a beautiful friendship."

We clinked flutes. The champagne was icy, dizzy, delicate stars down my throat. It was almost creamy. It was like drinking a 1920s black-and-white movie of people dressed in satin and ermine and black tuxes, being gay (the old-fashioned way) on New Year's Eve. The Snob reached across the napkins' matching heavy white linen tablecloth to hold my hands. I burst out laughing, not so much because of the presumed intimacy of the hand-holding or the fact that The Snob had just quoted *Casablanca*, the most over-quoted movie in cinematic history next to *The Wizard of Oz*. What was cracking me up was that this great-looking, beautifully dressed guy, a perfect stranger, was prepared to spend what I esti-mated to be at least $300 of his own money on dinner for a blind date. I'm not saying I'd have preferred Taco Bell—although Taco Bell has its place, usually on the turnpike—but between The Taco and The Forge there's a lot of middle ground. On the other hand, The Snob's predecessor was a married guy who only brought food in and was scared to be seen with me in public.

"Would you excuse me for a minute?" I said. "I need to use the ladies' room."

I wanted a cigarette and smoking was forbidden in The Vaunted Wine Cellar. I hadn't yet told The Snob I was a smoker. It would probably ruin my chances. The guy's a doctor specializing in children's *cancers*. Still, Argentineans consider themselves Eu-ropean, and Europeans all smoke. Fuck it. I'll tell him I'm a smoker—later. I dropped my finished Parliament into the toilet, flushed, powdered my T-zone, reapplied my Gladiator, and re-turned to the table. There were two little plates of pasta on it, and two crystal goblets of white wine. The Snob pulled out my green velvet chair. It looked like a tuffet, not that I'm sure what a tuffet is, exactly. Didn't Little Miss Muffet sit on one, eating . . . oh, what? Curds and whey. What *is* whey? *Along came a spider who sat down beside her and frightened Miss Muffet away.*

"I hope you don't mind, *ché*," The Snob said. "I ordered for us."

"No, not at all," I lied. Now was probably not the time to en-lighten this fabulous foreigner about American blind date women preferring to choose their own food, thank you. We get to pursue our own happiness in this country. It's guaranteed in our—

"Lobster ravioli with pink vodka sauce and a very nice, rich Chardonnay. Chalone Vineyards. California, but still quite good."

"Great choice," I admitted. "Great meal."

"Meal? This isn't our meal. This is our appetizer." Thank God I'd gone for a skirt with an elasticized waist. It'd be ridiculous not to in a place like this when somebody else is paying. "So, you're a writer. *Washington Post.*"

"Yep. Well, aspiring. I'm an aspiring writer. Mostly perspiring. This ravioli's yummy, by the way. Thank you."

"*Ché,* you're all flushed," he said, stroking my cheek.

"Well, seafood. Champagne. Wine. I'm fine."

"And I love what you're wearing."

I'd decided to pair my brushed gold flats (though I was still hell-bent on getting the fantasy red ones) with a fitted ivory silk vest with an ivory lace overlay and pearl buttons (with only an ivory lace bra underneath), teardrop pearl earrings, several strings of pearls wrapped around my wrist like bracelets, and a very full, drapey, multilayered, off-white chiffon skirt that hit at mid-calf. At the last minute I'd blended the *Out of Africa*'tude with *Tout feu, tout flamme* (All Fire, All Flame), a forgettable—except for the outfits—French flick. There's an outdoor scene where Yves Montand is helping to steady Isabelle Adjani on a bicycle. He's wearing a suit and a white shirt with an open neck. She, lovely as ever, is all in white: Loose ribbed crew-neck long-sleeved pullover, ankle-length and very full gauzy white ruffled skirt with white-on-white embroidery all across the tiered bottom half, and soft pale low-cut round-toed ballerina flats.

"Look at these women," The Snob whispered, gesturing across the room, the golden rectangular frame of his wristwatch catching the candles' and chandeliers' light. "They're so garish. Vulgar. Like gangsters' girlfriends."

"Goomahs," I said. I knew that word way before I ever watched *The Sopranos*. My "bi-talian stallion" friend Billy taught it to me.

"And those skyscraper heels, they're practically pornographic."

"It's Miami Beach," I said, shrugging. "Here it's vulgar not to be vulgar."

"Then we must be very vulgar."

"We are," I said. "We're *vulgarísimo*."

The rest of our dinner wasn't, unless you consider a bacchanalian banquet offensive. It was so unforgettable that after our date I wrote it all down in my diary: A whole *bottle* of Catena Alta, an aged Cabernet Sauvignon from my handsome escort's homeland; filet mignon with Grand Marnier *au poivre* sauce ("We Argentines are carnivores. Nothing's better than our *bife a la parrilla* [traditional grilled beef], but we'll make do"); potato skins with crème fraîche and Russian Beluga caviar; and arugula salad with shaved Parmesan, fennel, zucchini flowers, and roasted peaches—*roasted peaches!*—dressed in extra virgin olive oil and fresh lemon juice.

Don't cry for *me*, Argentina. I've got an elasticized waist. Also, I'm pacing myself and I *love* high-class doggie bags.

"Dessert menu, *por favor*," I told our dumbstruck waiter.

"Women don't eat anymore," The Snob said, smiling and finishing his wine. "Not in your country. This obsession with Atkins and cabbage soup and—"

"This is your country, too," I said. "This land was made for you and me."

"Well, maybe," he said. "I'm not sure yet. It depends on several vital factors."

"Such as?"

"Let's look at the menu, *ché*. What do you like?"

"Chocolate!"

I was torn between the praline-plugged chocolate velvet cupcake, and the "To Faint From," a crispy pastry shell stuffed with Heath Bar Crunch ice cream, drizzled with melted Swiss chocolate—and I'm never torn. If I'm torn, it's either because I want both (yes!) or I want neither. I'm very decisive when it comes to food and fashion and men. Especially fashion. I can scan a roomful of pretty dresses and pick out the best one in one minute flat. Thirty seconds if it's a drapey coral jersey goddess dress from Anthropologie. Nineteen seconds if it's a clingy yet fluttery cotton knit ivory lace slip dress with a clingy yet fluttery underslip from H&M. Five seconds for a fitted black bouclé racerback A-line dress by Isaac Mizrahi for Target. Excuse me, *Tarjay*. If I hesitate over a dress, then I know it's not The One. Same thing with a man. If I hesitate, then he's not The One. On the other hand, it is possible that men and love are more complex than desserts and dresses.

Or maybe not.

Because I have always believed that finding the perfect and elusive dress/shoes/red lipstick will make the perfect and elusive man I love love me more and want to marry me. But men don't care about my outfits and my makeup. At least not the way I do. So I've come up with ten basic observations:

1. See it.

2. Love it.

3. Charrrge it.

4. He'll marry me once he sees me in it.

5. Uh-oh.

6. Did the boy just LEAVE?

7. He's not coming back. Pass the Puffs Plus.

8. I'm actually starting to feel less Puffs Plus-y. I think
   I may be getting post-Puffs Plus. Those jeweled
   green suede moccasins with the tiny mirrors on top
   are adorable. They wouldn't go with a thing. I must
   have them!

9. Who? Paul/Arthur/Raymond/Steve/Rob *who*?
   These all, of course, lead to . . .

10. God, I love my wardrobe.

The Snob and I took the only reasonable path and ordered both chocolate desserts, plus espresso *and* cappuccino. As long as he didn't suggest a tango, I'd probably process the nine billion calories I'd somehow ingested over the past three hours without exploding like a piñata.

"You look like you could be a guest star on *Miami Vice*," I told him. We were enjoying a postbanquet stroll along the strand of South Beach. (Tangoing, no way. Strolling, okay-fine.) A full white moon shone down brightly from a black velvet sky. *It's only a paper moon, sailing over a cardboard sea . . . It's a Barnum and Bailey world, just as phony as it can be . . .* We were holding our shoes as the breeze tossed our hair and the tide pulled in, swirling around our ankles and back out to sea, the sand giving way underfoot. Salt water spray is not great on chiffon skirts or, in The Snob's case, suede loafers. But you know, there are times when it just doesn't matter. It's the seaquatic-perfect first date equivalent of Goddessina's "the kind of sex where you don't care if something rips."

"Oh really?" The Snob said, laughing. "Who would I be on *Miami Vice?*"

"A dark, foreign stranger. Someone very mysterious. And dressed as cool as Sonny Crockett."

"What about Ricardo Tubbs?"

"No. Sonny."

"Yeah, *ché*, but I don't smoke."

"I do." Shit. No time to censor the truth. So much for Delta housewifey idylls.

The Snob stopped strolling.

"You do?" he said.

"Uh, sometimes." *Good* lie!

"You don't seem like it. You don't smell like it."

"I know. I'm a tasteful smoker. I actually hate other smokers. Well, I hate other smokers' smoke."

The Snob kissed me. Stars. Champagne. Moonlight. The ocean. My blind date was officially fabulous. (And thank God he was old school; the last thing I wanted to do after a spread like that was . . . spread.)

"Let's sit," he said. "Smoke a cigarette." He took off his gorgeous leather jacket and laid it on the sand, leather side down. "For you, mademoiselle."

"You're too much," I said. "You sure it's okay? It'll get all sandy."

"It's only a jacket, *ché*," he said. "Besides, it'll make it smell like you."

Whoa. He wasn't kidding about Argentineans being carnivores.

Nothing like soft leather under a gal's ass to make her feel happy. I pulled my Parliaments and a lighter out of my tiny pearly purse. The Snob took the lighter and lit my cigarette. I blissfully exhaled an overdue stream of smoke into the salty, balmy night and leaned into The Snob's embrace. It was all delicious and I felt peaceful and pretty for the first time in a long time. Even with my lovely Gladiator kind of kissed away. Oh, well. Nothing gold can stay. (Except for gold ballet slippers.)

Part Two: MEMPHIS

"Flowers!" Goddessina announced. We were in the *Post*'s newsroom, me answering phones, she sorting mail and handing out FedEx and UPS packages and a transparent violet vase of cream-colored trumpet calla lilies. The attached note read: "Missing you, *ché*, with love."

"We love him," Goddessina said. "We love the deeply attentive, deeply pocketed ones."

"We do," I said, inhaling the decorative lilies' fragrance. "I still need to find my tomato red ballet slippers."

"Ooh, for your trip down to Elvis country. When is that happening, exactly?"

Since that perfect night in South Beach, The Snob continued the perfection over the long haul, or at least over the long distance between us. Flowers every week, phone calls every day and every night, for nearly two months. He'd repeatedly invited me to come visit him in Memphis. I kept stalling, blaming it on work. The truth was, I hadn't yet found my mythical slippers, those magical red ballet flats I was obsessed by and convinced I needed to seal the deal, as it were. I'd gone almost everywhere, too—Hecht's, Nordstrom, Neiman Marcus, Bloomingdale's, both Taylors (Ann and Lord &), Pappagallo, Talbots, a bunch of tiny shoe stores in Georgetown—and *nada*.

"You're crazy, dear," Goddessina said. "You don't *need* anything. You've got three thousand pairs of shoes. He's into you! Flowers. Phone calls. That amazing first date. First-class airline tickets to Memphis. What more proof do you need? Or do you just not like him?"

"I do!" I said. "I like him."

"Then what's the problem? Is it because you'd rather die than live in Dixie?"

"Here," I said, shoving a *Vogue* in front of her.

"Oh my God! That's your outfit!"

Almost. It was a photo of three models in acrobatically whimsical poses, all dressed in a slightly different version of the CUTEST Norma Kamali jumpsuit. It was fitted, in lycra and cotton, in a very graphic, fun, modern, bold black and white buffalo plaid. The models wore it with wide red hair bands, wide red belts, red lips and nails, and RED BALLET SLIPPERS. Buh-bye, Meryl Streep in Kenya. Hello, Diane Keaton in New York City. Told you things would change by fall. From all whites and ivories and pearls to red, black, and white together, with diamond studs.

In other words, I wanted to be Norma Kamali. I love women who are characters, originals, women who don't remind you of anybody else. *Vogue* had obviously copied ME. I'd seen yet another version of that Kamali jumpsuit in a mail order catalog before my Miami trip and had ordered that baby *tout de suite*. Mine wasn't even technically a jumpsuit; it was two pieces. (Can anybody really deal with a jumpsuit? Having to take the whole thing off just to go to the bathroom? As many TaBs a day as I drink? Please.) The top was a pullover with a jewel neckline and three-quarter sleeves. The bottoms were pull-on capris with slanted interior pockets. Even without a red belt (and I had no intentions of belting it or of doing the red hair band), isn't that the cutest you have *ever?* And it was so comfortable. Fitted, but not to the point that you couldn't overeat in it. Forgiving. So. That was my primary Memphis outfit, and we could not have liftoff *sans* red ballet slippers.

"Try Saks," my personal shopper friend Mindy suggested. "And if you see them there, buy two. I tell everybody to do that with everything they love."

Unbelievable. Like a dream, like a really really good dream, like God was watching over me, wanting me to follow Mindy's advice, I stepped Holly Golightly into Saks Fifth Avenue's own brand of red ballet slippers. With quilted gold tips. And grosgrain ribbon detailing. Made in Italy. $94. When $94 actually meant a lot of money. Poor Mr. Visa. *Quel* scream: I was a freakin' copy aide—don't let that "e" at the end fool you into thinking I could afford this—buying $94 shoes. Well, actually I was a freakin' copy aide charging $282, before tax. (Two pairs of red ones and an ADORABLE pair in the same style in *café con leche* leather— always good to have.)

Hooray! Hooray! Dixie Land! Land of cotton where old times are not forgotten! Prepare for yet another Yankee takeover! I'll take my stand—literally—in red ballet flats. *Look away, look away, look away, Dixie Land!*

I packed Paul Simon's *Graceland* and Joni Mitchell's *Hejira* audiocassettes for my Walkman. In a bouncy yet measured ditty, Simon sang about following the river down the highway of the cradle of the Civil War to Graceland: *For reasons I cannot explain, there's some part of me wants to see Graceland . . . I've reason to believe, we all will be received in Graceland.* Joni sang hauntingly about the ghosts of the "darktown society" of Memphis's old Beale Street, birthplace of the blues. She sang of the blues' father, trumpeter W.C. Handy. There's a little park there for him. Handy stands cast in bronze with a trumpet in his hand, *like he's listening back to the good old bands, and the click of high heeled shoes . . ."*

Or in my case, the tap of red ballet slippers.

The Snob met my plane at Memphis International Airport on a late Friday afternoon, bomber jacket-less. He ardently scooped me up into his arms (I live on Lean Cuisines during the week in order to be scooped up on the weekends) and kissed me the American way, informally, on the lips. Like that Southern hospitality. And love those gift shops. I kept getting distracted by all the deliciously kitschy tchotchkes. There was this one Elvis snow globe of Christmas at Graceland with swirling snow *and* silvery tinsel I really had to have for my otherwise comprehensive collection. I *love* snow globes. And oh, the refrigerator magnet possibilities! I had to have the hot pink Warhol-esque ELVIS '69 movie poster magnet from *The Trouble with Girls.* I'd buy them both on the way home.

It was still warm enough in early November for air-conditioning; The Snob immediately turned it on typhoon strength in his black BMW. I stretched my legs and gazed at my red shoes.

"Finalmente, ché," The Snob said, holding my hand. "My darling, darling girl. Thought we'd go home first so you can relax a little, unpack, and then—please tell me you like barbecue."

"I'm still a carnivore; I love barbecue," I said, squeezing his hand. It felt like it'd been dipped in Downy. I tapped my red ballet slippers on the gray carpeting. You couldn't hear a thing. Pile was too plush. "I like your suit and that snazzy tie, too," I added. "Guess you just came from the hospital?"

"Something like that," The Snob said. "I've had plenty of time to plan this."

"I know. I'm sorry. I've been . . . delinquent."

"Bad girl. You kept me waiting and waiting. But, working girls."

The Snob's place was so new I could almost smell the paint. It was a big town house with three bedrooms and three baths. Clean, spare, masculine, well lit, with touches of home: A framed photograph of Iguazu Falls, a volume of collected fiction by Jorge Luis Borges called *Ficciones*, a little clay gaucho hat with loose change in the brim. My guest room was great, with fresh red tulips in a silver vase on the dresser and my own private bathroom. I decided not to change—I'd worn a red silk and cotton twinset and jeans on the plane, and that had to be good enough for Corky's, THE place for Memphis 'cue, according to The Snob and that bitchy Queen Cruella from the *Post*'s food section. (Tomorrow at Graceland would be the coming out for the full-throttle Norma Kamali 'tude. I had to pace myself, just like I did at The Forge in Miami. Surely The King, blue suede shoes notwithstanding, would have understood.)

Everything was so orgasmically good at Corky's that I just wanted to put my face in it and slap that table with my red Saks flats! Have mercy! Racks of perfectly cooked moist ribs, sweet potato fries, coleslaw, and brown sugar baked beans, all washed down with gallons of iced sweet tea. Oh, and pecan pie and fudge pie and coffee for dessert. I could live here. Totally.

We rolled home—I could've used a wheelbarrow—and made out a little, just enough to intrigue me, and retired to our respective lairs. I needed my beauty sleep for Graceland. Wonder if they sell deep-fried peanut butter-banana sandwiches there?

✦ ✦ ✦

Norma Kamali at Graceland. It's a concept. The Snob had complimented my outfit at breakfast. Over strong coffee and bakery-bought beignets buried in powdered sugar, he'd said, "Wow. High fashion comes to Memphis. I never see it here, except on me. You look charming."

He stroked some stray sugar off my nose. I felt great. I didn't think I could feel any better with my Argentinean beau. *Delta Dawn, what's that flower you have on?* The sun was shining, my belly was full of sugar and coffee, my Gladiator was gladiating, my outfit and shoes were groovy, and The Snob was into all of it. As we drove to Graceland, I thought, This guy really gets me. Maybe Reizl—I mean, Graciela—was right all along. *It'll be like coming home. You two will have the cutest babies!* I was already imagining adorable tiny outfits and picking out pumpkins at Halloween time with them and taking pictures and how I'd get Corky's to cater their birthdays . . .

I heard The Snob snicker.

"Jesus," he whispered. "Could these crybaby rednecks BE any porkier? Or more pathetic?"

We and five hundred other people were clustered in the Meditation Garden behind Graceland mansion, where Elvis, his parents Vernon and Gladys, and his paternal grandmother Minnie Mae are buried. A white statue of Jesus oversaw them, and thousands of helium balloons, stuffed teddy bears, and real and fake flower arrangements threatened to obscure the markers. All you could hear were cameras clicking, water gurgling from the fountain in the little round pool beyond, mockingbirds twittering, and people crying.

"Check out that heifer," The Snob snorted, shaking his head.

He chortled into his Downy palm at a big fat lady who was loudly sobbing and blowing her nose into a tissue. She looked as bereaved as I am when a pair of my shoes dies. The lady was wearing a pink plastic visor cap with a black lightning bolt and the letters TCB on it, a reference to Elvis's famous slogan, "Taking Care of Business in a Flash"; a pink polyester short-sleeved T-shirt with a young Elvis's face on the front and ELVIS THE KING LIVES 4-EVER! on the back; pink sweats with twin black lightning bolts on both cheeks; a silvery ankle bracelet with what appeared to be a yellow ceramic hound dog charm dangling from it; a pink acrylic I ♥ ELVIS tote bag with the ♥ in pink sequins; and black rubber flip-flops patterned with tiny pink Elvises warbling into tiny pink microphones.

This was *haute couture* in Graceland. Criticizing that ensemble and its wearer would be like criticizing a cat for shedding on your cashmere sweater.

"You're being an asshole," I told The Snob. God, it felt good to curse. I hadn't cursed with this fucker before. And his permanent new nickname was *juuust* about to be inaugurated. "And you're a snob. You don't get it. You don't get the . . . *Elvis-osity* of it all. You're an anti-Elvis snob. THE anti-Elvis snob."

"I *don't* get it," he said. "Why are you angry?"

"I'll meet you back at the Heartbreak Hotel."

Really. There is a Heartbreak Hotel. It's across the street from Graceland. It's got a Burning Love Suite and everything.

"Come on," The Anti-Elvis Snob said, walking up behind me. "These people are hillbillies. Hicks. Hayseeds."

"UH-ppreciate the UH-lliteration," I said.

We drove in silence to the historic Peabody Hotel in downtown Memphis for lunch. I lit a cigarette in the car and didn't open the window—something I don't do even in my *own* car—expressly because I knew he'd hate it. He didn't say anything, though. We sat in the Peabody's drop-dead-gorgeous Lobby Bar, famous for its residents: live mallard ducks. They march in single file, and they're very friendly ducks who like to splash and quack in the bar's marble fountain. I wished I could've joined them. They looked like they were having fun. But the South was turning me into a piglet, not a fowl.

I started with a Bloody Mary—I figured I would need a lot of drinks, as I was OSP (Officially Still Pissed)—and a bowl of rosemary-tossed roasted cashews. Then I devoured a Peabody Club sandwich, a plate of fries with a gallon of ketchup, three buckets of Diet Coke (no restaurant on the planet serves TaB), and an Elvis-sized wedge of Equinox cake: "Hazelnut Bavarian with chocolate mousse, served over cream chocolate cake and hazelnut croquant." Whatever that croquant was, it was sensational. *Look away, look away, look away, Dixie Land!*

"Good appetite, *ché*," The Snob said, recoiling a bit. He finished his Heineken and pushed away his plate. There was still a portion of a Beale Street Burger and two and a half onion rings on it. I damn near ate those, too. "I've never seen a woman eat like Pacman and stay so petite."

"It's a mystery," I said, lighting a Parliament and sipping coffee. Like I was really going to tell Juan Perón over here in his silk shirt and dress pants and alligator belt and no socks and goddamn Gucci loafers that I live on meat lasagna Lean Cuisines during the week!

"Let's go," he said, getting up. "We can walk off lunch at the Brooks Museum. They have some great decorative European pieces and a few very good American paintings. You're an art history person; I think you'll enjoy it."

"I think I will, too," I said. "But not today. Next trip. I'm pooped."

"Come on," he said, playing the forceful alpha male to my would-be Delta wife. He pulled me up by the hands and began leading me toward the Peabody's grand entrance. "You can rest up in the car. They close at five. Then we'll go to Beale Street for drinks and hear some live music and go dancing."

I get it. This would be the Nazi death march—or should I say the Sherman death march—weekend across the Mississippi Delta. Also known as the endless activity date, designed to exhaust and entertain you to the point of intimate acquaintance with a city instead of your date who's accompanying you. It was like *Love Connection* with Chuck Woolery. My friend Billy thought the show was preposterous and bizarro, hence, he forced me to watch it with him sometimes. Must be a gay thing. Maybe Billy had the hots for Chuck. Every female contestant sure seemed to.

"These dates on this show go on and on forever!" Billy would always scream. "In the morning, the guy shows up with flowers and candy. Then they go to Bel-Air or Carmel. Then they go skating. Then out for dinner to a great restaurant. Then to Cancún. Enough already!"

"I need a nap," I told The Snob. "The King's jungle room kind of did me in. That green shag carpeting on the ceiling; I don't know."

"Come on."

"Or maybe it was the 'Pink Cadillac Garage' museum."

"Come on."

"Or the custom *Lisa Marie* jet plane."

"Come on."

"Or the 'Sincerely Elvis' museum."

"Come on."

"Or the *Walk a Mile in My Shoes* movie of Elvis's life. It's hard to say. But I've walked way more than a mile in my own shoes—Graceland's fourteen acres!—and the three of us are walked out."

"We don't want to miss the museum," he said, sidestepping a gaggle of green-headed ducks. Is it *gaggle?* Or is that just for geese? Duck . . . duck . . . goose!

"I'll meet you out front," I said. "I have to go to the bathroom."

My period had just started. Ohhh. THAT's why I'd been so starving. And snippy. I thought it was because The Snob had made fun of the people at Graceland like he'd made fun of the women at The Forge. Or because he wouldn't let me go home to take a nap. You know how obnoxious and tantrummy kids get when they're overtired and it's way past nappy time? Hormones. They rule everything. Thank God I brought along my little bag o' tampons and panty liners. Wouldn't want to screw up my brand-new Norma.

Breathe.

Powder.

More Gladiator.

Okay.

Fucking museum.

"*Did you* have a favorite?" The Snob screamed. "At the Brooks?"

I could barely hear him over James Govan & the Boogie Blues Band. We were sitting in the packed Rum Boogie Cafe, one of Beale Street's oldest bars. We were both drinking a strong, sweet, rummy concoction called Blueberry Hill that contained no actual blueberries. I was waiting for the Bacardi Silver to kick in so I could find my thrill. (Otherwise, I'd go buy a dashboard voodoo doll up the bustling street at Tater Red's Lucky Mojo's & Voodoo Healings.)

"What did I like best at the museum?" I screamed back at The Snob. "Well, it's hard to go wrong with Homer."

"You want to go home?"

"What? No. I mean, yes. I meant Winslow . . . Oh never mind."

I was hoarse and my feet were vibrating. Not from the R&B, but from the fatigue of the day. Looong day's journey into night. My little red ballet slippers had been beyond broken in, poor babies. Thank GOD I'd bought that second pair.

"Could we please go?" I screamed. I was ravenous. Again. Something deep-fried and cheesy . . . Maybe some deep-fried catfish with deep-fried Crisco to dip it in and some mashed potatoes deep-fried in cheddar cheese . . .

"*Ché*, I haven't finished my drink!" The Snob screamed. I wanted to grab one of the 350 autographed guitars mounted on the wall and just smash him over the head with it.

"That's your third one! But who's counting? Are you gonna be okay to drive?"

"You want to eat?" he said, slamming his glass down. "You'll eat."

"Deep-fried," I said. "Make it deep-fried."

I devoured a disgustingly delicious deep-fried double cheeseburger at Dyer's, where they even deep-fry the buns. So tasty! For a menstrual gal, this was heaven. I had my greasy burger with a platter of chili cheese fries (don't ask) and a chocolate milkshake. The South was rising again. In my stomach.

"Oh my God, I'm beyond stuffed," I said. "That was so good. Time for bed."

"We haven't hit one club," The Snob said. "They're just getting started."

"Go for it," I said. "I can get a cab home."

"What are you talking about, 'I'll get a cab home'? How would that look? We're going clubbing. I want to drink more and dance with you."

"My flight leaves early in the morning. I need some sleep. Good night."

He followed me out to the street as I searched for a taxi.

"Stop it," he said.

"You don't have to pay," I said. "I have cab fare."

"Okay. We'll go home. God. Where'd we park?"

Oh, that inspired confidence. Three Blueberry Hills and *Where'd we park*? Great. I should've smashed his head in with that Aerosmith autographed guitar when I had the chance; he'd never have felt the difference.

Just as I was thanking the Lord Jesus for getting me and my little red ballet slippers home intact and for finally putting an end to our fifteen-hour *Love Connection* death date, there was a soft knock at the door.

Fuck me.

"Gigi? Are you asleep?"

"Yes."

"*Ché*, I can see the light." Fuck me hard. "Can I talk to you for a minute? Please?"

"Go to sleep."

"Please?"

Fuck. Fuckfuckfuckfuckfuckfuckfuck.

I put on my glasses and a kimono and opened the door a crack.

"I'm really sorry, *ché*," he said. "I know you're upset."

"I'm too tired to be upset," I said. "What time is it?"

"Could we talk? Please? Come sit with me on my bed. I just want to talk to you. Let's not leave things like this."

I'd probably get to sleep sooner if I let him talk to me, so I let him talk me into it. His bedroom was like a Hollywood set. *Lost Horizon.* Emerald City. It had an Art Deco-Streamline Moderne feel, with dark walls and a huge silver sunburst mirror above the headboard. The king-sized bed was geometric, all inlaid wood. I settled on the comforter. It was brown and black zigzag, like a chevron design. The only light came from a mushroom-shaped frosted white lamp on the lacquered black nightstand.

"Interesting lamp," I said. Its base was etched with a profile of a kneeling naked archer ready to release an arrow pointed northeastward.

"Thank you, it's Lalique," The Snob said, sitting across from me. He was wearing nothing but a pair of cotton drawstring pajama bottoms he'd probably seen in *GQ.* They were pinstriped in gray and white herringbone. It was the first time I'd really seen The Snob's body. His hairless skin was supple and smooth, and he obviously worked out.

"Your pj's match your room," I said. "You are seriously coordinated. So glad I wore a Mexican kimono."

"What, it's adorable," he said.

"We clash."

"Look, *ché*, I know I've been overdoing it on our dates," he said, holding my hands.

"It's okay," I said. At this point I'd forgive him anything. Those pecs!

"There's a method to my madness, though."

He leaned in and kissed me. Damn, he was a good kisser.

Yowza. We were right back on Miami Beach that night. I knew nothing would happen—I was in major periodical mode—but maybe The Snob was right after all. Maybe this moment was the real beginning of a beautiful friendship. Everything softened. My attitude. His black curls. We lay down on our sides, facing each other in the Lalique lamplight.

"You know," The Snob said, stroking my arm with a fingertip, "in Freudian theory, the woman's vagina is lined with a double row of sharp, long teeth. Like a shark's mouth. And when the man inserts his penis, the teeth masticate it."

HUH?

I didn't know whether to cry, scream, laugh, or bite him. With my vagina. *Along came a spider who sat down beside her and frightened Miss Muffet away.* Oh God, was he Ray Milland and I Grace Kelly, and was this *Dial M for Mental Mastication?* This wasn't exactly the emerald cut Goddessina and I ever had in mind. I got up and backed away slowly from the massive bed.

"Are you really a doctor?" I asked. I don't why, it just blurted out of me.

"What?"

"Because your beeper hasn't gone off once. My father's a physician and even when he's not on call, his beeper always stays on."

"You're funny, *ché*."

"And you never mention your patients. I haven't even seen where you work."

"We passed St. Jude. On the way to—"

"Anybody can pass St. Jude. Elvis Presley's ghost can pass St. Jude in a pink Cadillac. That doesn't mean he works there."

"You want to eat barbecue and then go see sick kids in a hospital?"

"St. Jude. Isn't he the patron saint of lost causes?"

"All right, all right," he said. "Cards on the table. I'm gay. Satisfied? I'm a faggot. A queer. A closet queen. *Un maricón.* Nothing worse you can be than me, especially in the Latin world. No one knows. Not my family, not our friends."

"Graciela and Jorge—"

"Have no idea. And I'd thank you not to enlighten them."

I asked what I was supposed to tell them. Graciela was already naming our future babies (Juan and Liberace).

"She thinks you're *oro puro*," I said. "A golden boy." Fool's gold was more like it. I should always listen to dead poets like Robert Frost when it comes to human nature:

> *Then leaf subsides to leaf.*
> *So Eden sank to grief,*
> *So dawn goes down to day.*
> *Nothing gold can stay.*

"Tell them anything you want," The Snob said. "Just don't tell them the truth."

"Okay," I said. "I'll blame it on . . . chemistry."

"Don't. I need a green card. You'd make a perfect bride."

"Bride? You mean beard."

"Beard, bride, what's the difference? You'd want for nothing. You can have affairs, as long as you're discreet."

"Bait and switch," I said, walking to the door. "You know that expression? 'Cause I can't think of a better argument for legalizing gay marriage. Jesus."

"Will you think about it? Just consider it."

"*Buenas noches.*"

I closed the door softly behind me and returned to my room.

My red slippers were neatly placed just where I'd left them, on the floor by my suitcase. I stepped into them, closed my eyes, and tapped my heels together three times.

Dorothy was right.

And she *loved* red shoes.

Mother-of-Pearl Earrings

Every woman should have one tortured, sad poet in her past. It's like a romantic prerequisite. You astound him with your English Major fabulosity, verve, and youth, and he'll kill you (softly) with his Older Man scorchy love poems, rainy day moods, and unBE-LIEVABLE orgasms. Independence Day spectaculars. Crunchy toe curlers. Momentary-loss-of-consciousness dazzlers.

Really.

Specifically: If you're going to have a melancholy poet with a dirty mind for a lover, and Leonard Cohen is unavailable, you have to go with a cowboy poet. I used to tease mine about it because he was about as much of a real cowboy as Ralph Lauren. He was literary and libidinous and bohemian. He loved anything having to do with sex, the naughtier the better. And he loved the wide-open West, loved turning its nothingness into somethingness, into poems. That Big Sky loneliness—what the author Gretel Ehrlich called "The Solace of Open Spaces," which is the title of her lyrical Western book—was his major theme. Well, that and the mysterious intricacies of the human heart. Among other organs.

✦ ✦ ✦

To support his poetry habit, my Cowboy Poet worked as an editor and polemicist. His left-leaning, nonconformist, anti-censorship positions on almost every issue of the day were conspicuous and rankled folks in his overwhelmingly Republican state, particularly ultraconservative politicos. And he *loved* having that effect, that *épater les bourgeois*, shocking the middle class. It was his rebellion against The Man, who was . . . well, anybody who disagreed with him. *What're you rebelling against, Johnny? Whaddaya got?* My cowboy's complicated personality had a scrappy-sexy tinge of bluster and fight. He was a sensitive, jealous, easily hurt New York City runaway who'd married unhappily and fled to the West to cleanse and renew his troubled inner-city soul—and never came back. Well, never came back to stay. Otherwise we'd never have met. With his soft salt-and-pepper beard, big glasses, and dimpled, lecherous laugh, he was an earthy intellectual, a throwback to the Beats, a lapsed Roman Catholic Allen Ginsberg howling at . . . hm. Confinement. He wouldn't even wear a watch. It confined his wrist.

He never howled at me, though, except in ecstasy. We met over the phone when he called my office one day. Consequently, we got to be phone pals—he had this really sensuous voice, perfect for a cowboy poet. Gravelly yet mellifluous. Its timbre was like Chris Isaak's in his haunting, aching love song, "Wicked Game": *The world was on fire and no one could save me but you. It's strange what desire will make foolish people do.* The gorgeous Herb Ritts-directed black-and-white music video of that song, with Helena Christensen as Chris Isaak's lover—oh my God. They were on a beach and she was teasing him with her translucent pale green eyes and falling black lace bra strap and—oh my God. Like that song's meaning, my Cowboy yearned, capable of being filled but never fulfilled. Over a couple of flirtatious, confessional, long-distance years we became virtual intimates; later, actual ones. He was a woman's man, which is not the same as a ladies' man. His feminine side was more developed and assured than any man's I've known. The key to his cool was his ability to listen and empathize without judgment. He was a *great* listener. Very supportive and engaged. And he was a *great* lover. The kind of lover who you know no one else will ever be as good as. (And maybe shouldn't be.) The kind of lover—what the hell, let's summon Leonard Cohen—in whom you found what you would always want again: sensuality, affection, intensity, tenderness, imagination, and make-you-blush eroticism.

And he smelled good. Clean, the way the air smells right before it snows. When his big, long body wrapped itself around mine, emanating heat, it made me feel tiny. Sleek. As though I could slip into his pocket or sit on the awning of his eyelashes. My therapist used to tell me, "It can't be the Fourth of July every time you're together. That's impossible." But I swear, it *was* fireworks. Every time. With Cancers, sex is always coupled with feelings, with emotion; they're a water sign ruled by the moon.

Cowboy Poet had occasion to come east several times a year—for poetry readings, to cover politics, or when he couldn't stand the Western version of a bagel anymore (a Thomas' English Muffin with a hole punched in the middle). He'd invite me to join him and I was always excited to go, imagining all my outfits in various scenarios, especially off me and on the floor. He represented what I aspired to be: A published grown-up with sex appeal.

When my Hollywood psychic friend Kimberly Berg would say of men, "Let them come to you, you're the one with the goods," it never registered. I always thought THEY were the ones with the goods, therefore I had to go after THEM. This, of course, is what Dr. Phil would call stinkin' thinkin'. I wonder if Dr. Phil loves *All About Eve* as much as I do. There's this wonderful scene—well, the whole movie's wonderful; I can practically recite it by heart—where Eve and Bill are alone in her dressing room. Eve's coming on to Bill, who's in love with Margo. Bill admits he's curious about Eve and Eve tells him to "find out." Then Bill says, "Only thing—what I go after, I want to go after. I don't want it to come after me." Eve gets weepy and Bill tells her, "Don't cry. Just score it as an incomplete forward pass."

That was in 1950. Have things really changed all that much? Is this why I'm still single? 'Cause based on results, I must be doing something wrong. Dr. Phil says men are biologically hard-wired to pursue. *I chase, therefore I am.* But if I say, Great! Come and get me, honey!—I'm doomed? I'm not supposed to let them know I like them, or else they'll take me for granted and lose interest?

But if I didn't choose to chase, I wouldn't have my little pink raincoat. Or my peach panties. Or my red ballet slippers. Or a pair of unbearably beautiful and absolutely unaffordable Ted Muehling mother-of-pearl drop earrings from Bergdorf Goodman. I get Bergdorf's catalog in the mail—I get every catalog—and gaze at all the gorgeous pictures for hours. They torture me because they denote a world that's out of reach for a poor girl with inconveniently expensive tastes. I'm telling you, having good taste is a curse. Anyway, in this one Bergdorf catalog, I saw Those Earrings. They jumped out at me, screaming my name. Those Earrings haunted me, even in my sleep. I dreamed about them the way I dreamed about my Cowboy—lustfully. When I couldn't take it anymore, I'd wake myself up. But then I'd fantasize about them awake, imagining wearing them with a fantastic outfit for my upcoming tryst.

I always wore pearls with my Cowboy Poet, but with the Bergdorf ones I'd be extra, extra devastating. My love would finally come to his senses, leave the freakin' Rockies, and return to the East for good, for *moi*. I just needed Those Earrings. Those Earrings! Those Earrings!

Two hundred twenty-five dollars.

Sigh.

GOOD GIGI: *Are you insane? You couldn't justify that purchase if you were high on X. You have a bulging drawer full of perfectly beautiful pearl earrings in every shape and size, every color. Go to sleep. Forget it. You're a big girl now. Being able to tell yourself no is a sign of—hello!—maturity. Your life will s omehow continue without $225 Ted Muehling mother-of-pearl drop earrings.*

BAD GIGI: See, everything you just said is so wrong that I hardly know where to begin. In the first place, I can live without the sarcasm.

GOOD GIGI: *You're right. Sorry.*

BAD GIGI: And I can live without "food" for a month. I'll go on my TaB and Parliaments diet. Low-fat, high-fiber.

GOOD GIGI: *(Stunned silence)*

BAD GIGI: If you can wear only one piece of jewelry, it should always be earrings. They finish you, they frame your face and complexion. And no female can ever have too many pearl earrings! I am the ORIGINAL *Girl with a Pearl Earring,* Johannes Vermeer portraits and Scarlett Johansson movies notwithstanding. My first pair of earrings were tiny pearls, and I was a baby! So I've got legitimate roots in this. What's more, a girl needs aspiration and Bergdorf Goodman catalogs are aspirational. They're dreams. I want Those Earrings.

GOOD GIGI: *I know how much you love those earrings. And yes, they're beautiful and you'll have them forever, so they're like investment shopping.*

BAD GIGI: Exactly! Thank you!

GOOD GIGI: *Hold it, missy. What about your Metropolitan Museum of Art ones? Aren't they almost identical? Be honest.*

BAD GIGI: They are SO not identical that I'm almost embarrassed to be having this conversation. The Ted Muehlings are convex circles. The Met earrings are teardrops. Totally different. And the Met ones are called Venus because they're adaptations of . . . wait, I'll get the Met copy from their catalog . . . "those worn by the Roman goddess of love in *Venus in Front of a Mirror,* a famous painting by Peter Paul Rubens (Flemish, 1577–1640). In the painting, Venus wears nothing but a bracelet and a pair of pearl teardrop earrings—one of the earrings is white, and the other, reflected in a mirror, is black. Produced in cooperation with the Princely Collections of Liechtenstein. 24 kt. gold plate, with handmade glass pearls. Length 1¼ in.; width ½ in. Pierced. Available with white pearls. One white pearl, one black pearl and black pearls not available. Member Price: $43.20 each; Non-Member Price: $48.00 each."

GOOD GIGI: *And your point is?*

BAD GIGI: *(Obdurate silence)*

GOOD GIGI: *Are you there?*

BAD GIGI: He's already seen me in them!

GOOD GIGI: *When?*

BAD GIGI: Last year. Remember? At the Maryland poetry reading? He wore that scary dashiki over a pair of old cords and I didn't say anything?

GOOD GIGI: *He wore a dashiki?!? I must've blocked that detail.*

BAD GIGI: And then he told me I looked every bit the Western bride? See? BRIDE. That means he's THINKING about it. *Another bride, another June, another sunny honeymoon . . .*

GOOD GIGI: *Even if he did say that, SO WHAT? They're pearl earrings! As if a guy would even notice which ones you were wearing at any given time! They can't even remember what they ate for breakfast this morning!*

BAD GIGI: That was such a good outfit. It took me forever to figure out. What do you wear to your first cowboy poetry reading without looking like Dale Evans?

GOOD GIGI: *A dashiki, maybe?*

BAD GIGI: It was a white long-sleeved pullover Norma Kamali fleece dress that felt like a sweatshirt, with a cinched waist and a very full long skirt with a deliberately asymmetrical hem; a white lace scarf around my neck; opaque white tights; pale pink cowgirl booties; the Met Venus earrings; and clusters of Carolee pearls everywhere else.

GOOD GIGI: *That was a good one, I'll give you that.*

BAD GIGI: And later that night I sat on the edge of the bed and he slowly pulled off my little pink boots. And I dropped back on the bedspread and closed my eyes and my heart was going like mad and I was Molly Bloom *yesyesyesIsaidyes* and all I heard was the click of his belt buckle as he unfastened it. So exciting. It was the sexiest thing that ever happened to me.

GOOD GIGI: *I'm so glad I'm not the real you.*

For this upcoming rendezvous, I wanted a different look. Still soft and feminine, of course, but with an edge. It was winter, I wanted black. My Cowboy was coming to town to accept a *Playboy* award. I'm not kidding. Every year since 1979, Playboy Enterprises, Inc., has presented the Hugh M. Hefner First Amendment Award to "individuals who have made significant contributions to protect and enhance First Amendment rights of Americans." People like Michael Moore, Bill Maher, and Penn & Teller have won it. I subtly suggested to the Cowboy Poet that he go for broke and buy himself a decent suit and tie, as he owned neither. His idea of dress-up, when the dashiki was in the wash, was an ancient tweed jacket with elbow patches, a button-down oxford shirt, navy cords, scuffed brown suede boots, and—I swear—an ascot. He owned an entire *wardrobe* of ascots. Granted, Cary Grant and Fred Astaire wore "neckerchiefs," as they were called then, and Cary and Fred were nothing if not stylish. And my cowboy did wear his ascots with a star's relaxed confidence. It was his individual expression. And who's more into individual expression than I? Hello. I respect it. I live it. However. An ascot is not a substitute for a proper necktie. It just isn't. Never will be.

"You're a *winner*," I told him, appealing to his ego. "That's so special! It's a big, big deal. An honor and a privilege. Winners wear beautiful suits and dress shirts and ties and shiny shoes. It's so sexy."

"It is?" he said.

"Totally. If I saw you dressed like that I'd jump on you."

"You jump on me anyway."

"I know. But I'd REALLY jump on you."

"Those duds cost a lot of money, kiddo," he said. "I just don't have it."

I told him I could overlook the suit and the dress shirt and the good shoes (love is, after all, about negotiation and compromise—what a pain in the ass—and you have to pick your battles) but the ascot had to GO.

"Please wear a tie, darling," I said sweetly. "For *me?*"

"No."

"I'll buy one for you," I said brightly. "It'll be an early Christmas present. Silk!"

"No."

"I'll let you tie me up with it afterward, honeybunch. What do you think?" Very *Basic Instinct.*

"Hmmm."

"Okay?"

"Sorry, sweetie. No. Besides, I've seen Hef wear ascots."

"Where? When did Hef wear one?"

"I don't remember. I just know he did. He has."

"Okay, well, I'm not debating whether Hugh Hefner is or isn't an ascot person. I'm just saying, I'm trying to figure out how to live on TaB and Parliaments for YOUR event. Won't you please, please suck it up for me this one time? It's really for YOU."

"Well . . ."

"Okay?"

"No."

"Why NOT??"

"Those earrings are YOUR trip, honey. Not mine."

You know what it is? I know what it is. It's that hard shell some of them have. Cancers. They're crabs. You can't tell them anything. He should've been a pair of ragged claws scuttling across the floors of silent seas. IN AN ASCOT. Bergdorf Goodman pearls before swine, I swear to God. Bergdorf Goodman pearls before crustaceans. Look: When a woman of obvious taste and style offers to help a Cowboy of obvious, uh, NONE, and he has the audacity to refuse—what can you do? Why, you call up Bergdorf Goodman and order Those Earrings, that's what you do. And while you're at it, you order that killer off-the-shoulder jet black mohair-cashmere pullover sweater, too. That'll teach the fucker—until the Visa bill arrives. Just because *he* thinks wearing an ascot makes him a "rebel" doesn't mean *I* have to follow suit. *I* rebel by buying earrings I cannot afford.

A week later, my Bergdorf Goodman packages arrived. I had to pop an Ativan just to calm myself enough to open the orgasmic silver gift boxes with the store's name and illustrated Fifth Avenue façade in fuchsia. First, the sweater. Tissue paper, tissue paper. Oooh. It was *good*. By a company called 525 America. Light as a zephyr, perfectly fitted, very please-touch-me texture, upper thigh-length, with a wide funnel neck that could be worn as was or rolled down past the shoulders. (I'd do the latter. Better to set off my shoulders and Those Earrings.)

Now for Those Earrings. Be still my beating heart. Oh my God, I'm in LOVE! They were thick, iridescent mother-of-pearl dangling oval chips, 2" long and ¾" wide, with delicate 10-karat yellow gold tops and open-backed 14-karat yellow gold wire loops. I put on a pair of cotton gloves—didn't want to mar any-thing—and tried them on. Wow. *I'd* freakin' marry me.

Time to try on the rest of my *Playboy* outfit. (No pink bunny ears or white cotton tails until AFTER the party.) Those Ear-rings. The 525 sweater. Worn *over* the ivory chiffon layered skirt

I'd worn to The Forge a thousand years ago. Remember? I love trend-transcendent, seasonless clothes. So enduring and versatile. I can remix them forever for new seasons and new men. Good clothes always outlast bad exes. The fundamental things sure *do* apply as time goes by: I can wear my ivory chiffon skirt to meet my poor girl's Leonard Cohen in Washington, D.C., and it'll be all news to him. Here's looking at *moi*, kiddo.

Where was I? Oh, yeah. Opaque black stockings. Black velvet booties with princess heels. Small black velvet drawstring (with little matte gold crosses on the ends) pouch purse with a twisted black silk rope strap and a dainty silk black tassel dangling from the bottom. No other accessories. Just Gladiator-rouged lips.

That's poetry.

He flew into town the night before the *Playboy* awards ceremony so we could spend some high-octane time together. This always seems to work exceptionally well inside a lavish hotel suite paid for by someone else. Thank you, *Playboy*. (That mini-bar alone. Wow.) After alienating our hallway neighbors and the hotel staff with our, how shall I put it, loud luvin', we donned the suite's pair of fluffy unisex terrycloth bathrobes and attacked the room service menu. Is there anything better than perfect food after perfect sex? No. Thank you, *Playboy*. The Poet and I shared almost everything, and this decadent dinner was as pornographically arousing as it gets. Truly smutty in its fabulosity. We began with a yummy Chardonnay from a Napa Valley winery called Cakebread. (Don't you just love that name? *Cakebread*. Like a little fairy tale.) It accompanied a plate of fist-sized mushrooms stuffed with jumbo lump crabmeat, a bowl of onion soup au gratin, and a Caesar salad. (Personally, I could've stopped there. But why change horses in midstream? I'd spent the previous two hours screaming, "Don't stop!") Then we had an amazing Beringer Cabernet Sauvignon, sincerely scandalous steaks with thick béarnaise sauce (my absolute favorite, I could've swallowed it straight from its tiny silver serving pitcher), grilled asparagus with lemon butter, and *pommes frites* (which I oh-so-unsophisticatedly slathered in ketchup). I finished the wine and took a much-needed Parliament break. The Chocolate Sin cake topped with broken shards of macadamia brittle awaited. Yee-ha! Crack that brittle, crack that crab's shell, get to the mushy part!

"We could be like this every day, you know," I said, exhaling my smoke. So good after a big, rich meal. "It could be our life."

"This?" he said, chuckling. "This is a *dream*, darling."

"No, I don't mean this," I said. "Not this-this. *This.* You and me."

He handed me my coffee. It was in a gilt-edged porcelain cup with a matching saucer. The white background was overlaid with an intricate design of interlaced blue, lavender, orange, and lime flowers, and golden garlands. The touches of gilt reminded me of Gustav Klimt's painting *The Kiss.*

"I mean having dinner and talking," I said, twisting my hair back and up with a tortoiseshell barrette. "Drinking wine and making love and . . . like this. See? Like the flowers on this cup. Intertwined."

I put my cup on the table, finished my cigarette, and turned the saucer upside-down. "William Yeoward," I said.

"What?" he said, sitting next to me on the bed's edge. He munched a brittle shard and put one in my mouth. Heinously tasty.

"The china. It says 'Made in England' by someone named William Yeoward."

"Hey, when do I get to see your new earrings?"

"Tomorrow night," I said, swirling my fingertip in the Chocolate Sin's bittersweet sauce and slipping it in my mouth. Heaven. "You can't see them now. It's bad luck."

"That's what they say about wedding gowns."

"Yes, groom," I said. "That is what they say."

"You have such pretty eyes. Luminous." He took me in his arms and kissed me deeply. I felt his kiss roll and slide all through me. He licked a straying streak of Sin off my chin. He took the saucer from my hand. He took my barrette off. He took my bathrobe off. He took my eyeglasses off. He stroked me all over, as smooth and warm as béarnaise and Sin.

"They're pretty but they don't see," I mumbled, closing my eyes. "Don't see, groom."

"What?" he whispered, bending me back on the bed.

"My eyes," I said with a sigh. "They don't see a groom."

"*He'll never* marry me," I told Jean-Paul, proffering a square of aluminum foil from the precut stack on my lap. It was the next morning. While my Cowboy Poet was getting his *Playboy* award over breakfast and schmoozing and having lunch and schmoozing some more, I went to get my hair done for the reception that evening. Color, highlights, cut, and blow-out. Close to four hours right there.

" 'Hill nahvir marry me'—pliiiz," Jean-Paul said.

"Yeah, but I got these new earrings and . . ."

Jean-Paul rolled his eyes. We'd been down this road so many times before that his Ermenegildo Zegna loafers were worn out.

"Vat else is nue?" he said.

His hard-to-place, vaguely continental accent was reminiscent of wedding planner Franck Eggelhoffer's in *The Father of the Bride* movies. Maybe this is how all the Armenian-Lebanese Christians talk? That's what Jean-Paul is. He snaked the end of a black rattail comb in and out of my long curly hair, snatched the foil square, and placed it under the selected strands. He brushed them with the thick white paste that would turn me into sun-kissed divinity, and neatly folded the lot in place with the rattail into a perfect smaller square. We would repeat the procedure numerous times before my hair was "naturally" highlighted. The refined results always justify the effort and dollars involved. Once you find a good hairdresser, never let him go—that's my motto. I've followed Jean-Paul Mardoian and his scissors, brushes, and pithy, epigrammatic life advice for years. I schlepped to the Watergate when he worked there, and now I schlep to Foxhall Square in Northwest D.C., where he has his own salon on New Mexico Avenue. Even when I lived in Raleigh, North Carolina, I went to Jean-Paul (273.23 miles ONE WAY). And here in New Jersey, too. Every month. (Love that E-Z Pass).

I sat under the dryer for my highlights to process, sipping coffee and reading *W.* I wondered if Helena Christensen had a Cowboy Poet, too, or someone like a Cowboy Poet, an unavailable—and therefore more desirable—love object. Helena Christensen could easily afford Ted Muehling mother-of-pearl drop earrings from Bergdorf Goodman. But Helena Christensen didn't *need* them to make a man be with her like I did. Because she was Helena Christensen. Cowboy Poet was with me but not like I wanted him to be: with me every day. Every night. Not just in the same zip code. In the same home. Entwined, like Chris Isaak and Helena in the *Wicked Game* music video. Like the flowers on the English coffee cup. Together, in our own bed. *Everything depends upon how near you sleep to me,* as Leonard Cohen sang. I wanted us to be the two sleeping lovers in Toulouse-Lautrec's wonderful painting *Le Lit* (The Bed). Facing each other, cozy and safe and peaceful, with the blankets up to our chins and our hair strewn across white pillows.

Fuck.

Those Earrings would have to be stardust.

I love hotels. I love everything about them. That sense of anticipation you feel when you first walk in, as if something new and exciting is just about to happen. That anonymous feeling you have in them, where even in your own city you feel like a traveler. And all that ice! *Quel* luxury! Those eerie silver ice machines down the hall—God, I love them. I love having the bathroom all to myself, with my five billion beauty products scattered across every inch of counter space and a lit Parliament in a nearby ashtray next to a fresh glass of TaB on the rocks (I always bring my own six-packs; no place serves TaB)—such *mises-en-scène* make me delirious.

My Cowboy had left a note under my bag of Gladiators and complementary glosses saying he'd pick me up at seven. It was 6:15. My hair was Jean-Paul'd to the max, lustrous and straight as a satin curtain (I have really curly hair but he looks down on curls, calling them "chip-looking," and always straightens them to within an inch of their lives). I'd pulled the front part back with a black velvet barrette with tiny black crystals on it so you could see my shoulders and Those Earrings. Makeup, fresh and perfect. I had four tasks left: First, spray on my new perfume. The Cowboy Poet had given me the floral, feminine Yves St. Laurent perfume Paris the night before as a sort of conciliatory gesture over The Ascot Business. He said Paris was "classy and soft and sweet, like you." (But he still wore his ascot.) Next, I had to dress (never want to do the makeup with your clothes on), put on Those Earrings (yay!), and apply my lipstick to achieve ultimate Gladiation. Cowboy Poet! Acquiesce to my Chris Isaak-Helena Christensen-Gustav Klimt-*All About Eve*-Leonard Cohen-Toulouse-Lautrec dreams! Right now!

Okay. Forty-five minutes left. It would be tight—perfect application of red lipstick requires incredible patience and precision—but I could probably just make it. I locked the bathroom door. I don't like men walking in on me until I'm completely ready to be unveiled. When I'm doing my beauty rituals, leave me alone. Shoo! I say, shoo!, with a wave of my powder brush. Out, out! Don't want to kill *all* the mystery.

Seven o'clock. I heard him amble in and call me.

"It's Hugh Hefner, honey," he said, laughing. "I've come to show you my ascot."

"Congratulations!" I said, tearing out of the bathroom. I jumped into his arms, careful not to mess up my makeup.

"Whoa, baby. I'm an old guy. Let me take a look at you."

"I'm so happy to see you!" I said, hugging him and not letting go. "How was it?"

"I'm happy to see you, too, darlin'. Jeez, your hair. It's all . . ."

"I know."

"And you smell beautiful."

"It's your Paris," I said. "Do I smell like home?" In one of his poems, he'd advised girls to "marry a man who smells like home."

He let me down gently and backed away, holding me at arm's length. He sighed.

"God, you look so beautiful," he said. "Turn around, let me see the whole thing. I don't think I should let you out of this room. Guys'll be surrounding you."

"Those Earrings workin' for ya?" I asked. I tilted my head so he could touch one.

"Everything's workin' for me," he said, nuzzling my naked neck and shoulders. "These are grand. All glowy next to your skin. Like little moons."

"Mother-of-pearl. Deep as cats' eyes."

"One little grain of sand, huh? Turns into pearl. Botticelli's *Venus*."

"Hey, you're the crab, Cancer man. You're who lives with the other mollusks a full fathom five."

"Those are pearls that are your eyes." We played with Shakespeare quotes; one of us would lead and the other had to follow, or try to.

"I have to start from the beginning," I said as we walked to the elevator arm-in-arm. In my velvet princess-heeled boots I could almost reach his tweedy shoulders. " 'Full fathom five thy father lies;/ . . . Those are pearls that were his eyes—' "

"Skipped a line there, lady."

"I did?"

" 'Of his bones . . . ' "

"Oh yeah," I said. " 'Of his bones are coral made;/Those are pearls that were his eyes;/Nothing of him that doth fade,/But doth suffer a sea-change/Into something rich and strange.' "

A room full of *Playboy* people and no Hef? It's positively perverted.

"Christie said her dad couldn't make it," my Cowboy said. "Tied up with something."

"To his bedpost with silk neckties, no doubt," I remarked.

Tuxedoed waiters wafted about, extending endless silver trays of Champagne and Martha Stewart-worthy hors d'oeuvres. So much more fun and varied than a conventional dinner. You ate whatever and sat wherever you wanted. While the Cowboy Poet went to get us Champagne and the ghost o' Hef hovered, I executed my Jewish survival strategy, which has served my people well for centuries: Skip the bullshit. Get the food.

By the end of Scavenger Hunt Round One, Dinner, I was so overloaded with red linen cocktail napkins full of fabulous finger foods—crab cakes with chili-lime aïoli, molasses-glazed cocktail ribs, Chinese Pearl Balls with scallion-soy dipping sauce—that I had to retreat to a table and drop them off like a dump truck before I could start Round Two, Desserts.

Thank God I had a date to blame the glut on to my table mates. Then again, it was dark in there, just tons of fat little candles, so you really couldn't see much beyond Those Earrings and my Gladiated lips. Apparently, they were something to see. A handsome young man from Playboy Enterprises in Chicago introduced himself and pulled out my chair. A handsome young man from a New York magazine offered to get me Champagne. A handsome young man from a Los Angeles newspaper asked if I was related to Michelle Pfeiffer. I felt just like Miz Scarlett at the barbecue, surrounded by the Tarleton twins plus one, all wearing beautiful suits and dress shirts and silk ties and shiny shoes. Thank you, *Playboy.*

New York magazine man and the Cowboy Poet showed up at the table at the same time. They both sort of looked at each other, each holding two bubbling flutes of Champagne.

Beat.

"Thank you!" I trilled. "Bring on the Champagne."

"Do you mind?" Cowboy said to New York, taking New York's seat, which was next to mine. "She's with me."

"Oh, no problem," New York said, handing me my flute and sitting in another chair. "I'm her Champagne manservant. That's my official title, I think."

"We love this woman," Playboy Enterprises man said. "She's something else."

"Yeah, I'm still not posing for you," I said, biting into my crab cake. So good! "No nudity in public."

"She's so refreshing!" L.A. newspaper man said. "We've been telling Gigi how she resembles Michelle Pfeiffer. Something around the mouth, the shape of her lips."

"Yes, and I've been telling them that they're very descriptive psychotics," I said.

"Can I talk to you for a second?" Cowboy Poet said, taking my arm.

"Save my seat, fellas," I told the table. "And don't eat all my Pearl Balls."

He escorted me to a far corner, cornering me into the corner. The Champagne was fizzy in me. I stroked his cheek. He took my hand and kissed my palm. Then he held me by the wrist. Not hard, but enough to make his point.

"I knew I shouldn't have let you out of our room," he said. I pulled my wrist out of his grasp, shook it, and finished my Champagne. "They all want to fuck you."

"So?"

"I just . . . maybe Those Earrings are working too well."

"On the wrong guys," I said. "I bought them for *you.*"

"Right, but they're working on everyone. And your sweater, you know, black next to bare white shoulders, and your hair . . . I'm going back to that goddamn table and give them all a knuckle sandwich."

I started laughing so hard I almost lost my velvet-heeled balance. *A knuckle sandwich?* So retro! I could see the headline now: *COWBOY POET BRAWLS AT PLAYBOY BASH!* Several guests turned their heads in our direction. I covered my Gladiated Michelle Pfeiffer mouth.

"I'm sorry," I said. "Did you just say 'knuckle sandwich'?"

"Yeah. Teach 'em about messing with my woman."

"Oh my God! We're in the wild, wild West now?"

"That's right."

"See, here's the thing, dear. We're not. We're in the East. It's relatively tame here, outside of the Bronx. And you know, Hef may be into ascots, but he's not into violence. Isn't he your role model?"

"I didn't like what I saw, that's all."

"Should I wear a burqa?" I asked. "Would that make you feel better?"

"I don't think they sell those at Bergdorf's."

"How would you know? Karl Lagerfeld could design a Chanel burqa. Black with white camellias on the front. That would be so chic! But I mean, you're so far away. How would you even know? If he did, I mean."

"Honey," he said, "you're losing me."

"You're losing *me*," I said. "You don't get to be jealous. After tonight you're going back home. And I'm going back home. And that's it. I love you but this is really ridiculous. You're not going to be with me. You said so yourself, that this is all just a fantasy."

"I said *dream*."

"I've offered to move a thousand times and each time you say no, that I wouldn't like it out there, that it'd be too lonely for me, or you say that I'm no cowgirl or some other lame excuse like that!"

"You *wouldn't* like it. Not to live. And I don't like it here, you know that. I've already lived in the East, it's not for me. It's too confining. Please let's not have another tussle about it."

Another *tussle. Another bride, another June* . . . The Champagne was wearing off. I needed more.

"Our Chinese Pearl Balls are getting cold," I said.

As I left my corner and walked back to our table, I touched my earrings. They felt cool and smooth and hard. More than enough to beguile a man. Hell, three men, in three nice suits.

But not a single Cowboy Poet in a solitary ascot.

5

Backless Black Dress

Miles and I were in the same Classics of the Modern Theater: Realism and After class at the University of Maryland in College Park. I couldn't stand him. Miles interrupted. Miles cross-examined. Miles showed off. He dominated the class by intellectually overpowering and intimidating the passive professor who stood there staring at Miles in a stupor of self-doubt.

And that's what Miles was doing today. It was as if it were Miles's class and Miles was teaching us Federico García Lorca's *Blood Wedding*. According to Miles, the reason everyone in the play was so wounded and miserable was because they were all closeted homosexuals and lesbians. This was also Miles's basic explanation for all the world's woes. I kept looking around the room; except for mine and Miles's, every face was inert, submissive as a sheep's. The silence of the lambs. And Miles, their garrulous gay shepherd. God. Some queens you just wanna stick back in the closet. Please! Get back in there and SHUT UP! But this one wouldn't. I saw that if I didn't act and soon, I, too, would become a Miles ruminant.

Act 2, Scene 1: A Lorca character named Leonardo says, "To burn with desire and keep quiet about it is the greatest punishment we can bring on ourselves . . . When things get that deep inside you there isn't anybody can change them."

Got it, Señor Lorca!

I abruptly got up—Miles always stood to better pontificate—and said, as much to Miles as to our professor, "Excuse me. This is bullshit! Who the hell is teaching this class?" I looked at the professor. His head was bowed. *Bowed.* "I know *you're* not. I mean, okay, this isn't exactly Harvard but my family and I are still paying good money for my tuition here." I turned to Miles. "And we're not paying it to hear YOU hold forth about any goddamn thing that occurs to you. So get off the bully pulpit, BE QUIET, and let's have the TEACHER teach the fucking class."

Silence.

Then one pair of hands in the back began clapping. Then another. And another. Soon the whole room erupted into whooping applause, Miles included. I sat down, more exasperated than flattered. It's exhausting to be the proactive mensch. My sheepish professor cleared his throat, adjusted his eyeglasses, took a deep breath, glanced at me gratefully, and proceeded to teach what was left of Lorca. To tell the truth, I half regretted what I did; Miles *was* a lot smarter and more entertaining than our teacher. But that wasn't the dispute. I saw Miles's mouth open to interject once or twice but each time I shot him a look equivalent to an electric cattle prod zapping him, as Dorothy Michaels said in *Tootsie*, in the badoobies. Miles learned fast. It was positively Pavlovian.

Later, as I was gathering my books and papers and returning students' high fives, I heard a meek, high voice say, "Wow. I'm so in awe of you."

Miles was standing there with a grin on his strange little New Wave Germanic face. Small, narrow eyes. Broad nose. Full lips. Plucked eyebrows. Was that purple mascara on his eyelashes? He was so fey he made me look like the Brawny paper towel man.

"Thanks," I said, hoisting my bag over my shoulder.

"No one's ever stood up to me like that," he said, following me down the hallway like a lovesick lamb. "You're so radical. I love that. Can I buy you a coffee, honey? Are you free? You should come over for dinner. You like Indonesian food?"

"*Miles tells* me you're a rabble-rouser and funny—I like that," the Israelite said in his heavy Hebrew accent. It made everything he said sound droll, dour, aggressive, categorical, cocky, opinionated, ironic. We were meeting each other for the first time *chez* Miles for dinner. Miles lived just off University Boulevard in Adelphi. What can I say about Adelphi? That it's gross? That it's squalid? That it's a total pit? That the only reasons you'd ever go there are because you got lost or you want cheap gas or you have an odd hankering for authentic *pupusas*? Well. Adelphi's got one thing that's so rhapsodically good you can overlook the cheesy setting: Ledo's. Ledo's pizza. This pizza could inspire you to arias. It's rectangular, greasy, rich, and just incredibly orgasmic. There are other Ledos around town, but they're not the same. You gotta hit the original for its unique 1955-era grease. (And don't wear anything that doesn't give around the waist.)

Miles's place was a dim, grim rented room with a tiny kitchen and bath in a dark old creepy house with an untended yard and a tethered one-eyed dog who looked really upset and never stopped barking. It was Southern Gothic, Boo Radley spooky. Following my "This is bullshit!" outburst the week before, Miles had told me about his Israelite crush over cardamom-ginger coffee and apple cinnamon couscous cake (not entirely disgusting, considering it was "health" food) at the Student Union's Maryland Food Co-op. The Israelite had recently moved to the States from Tel Aviv to get away from a person so horrible that he refused to name her, calling her Mademoiselle X, and to study filmmaking. That's how he and Miles met, in a History of the European Motion Picture class.

"I'm not a rabble-rouser," I told the Israelite, laughing at the way his accent made the phrase sound. "I just . . . I'm an assertive person."

"I'll say," Miles said. He handed me a plastic glass of cheap red wine and an unwashed aluminum ashtray. "You almost scared me straight, honey."

I doubted that was possible. Miles handed the Israelite a cold can of Schlitz and fondled the top of his head before gliding back to the kitchen as if on ice skates. The Israelite didn't respond to the gesture but he didn't seem to *not* like it. He lit my Parliament and his own Kent. The cigarettes' smoke smelled way better than whatever that appalling stench was emanating from the kitchen. Was it . . . burning curry? We seemed to be on the same wavelength; the Israelite pinched his nostrils and rolled his eyes. Maybe it was the Hebrew Homeland Connection; I felt we understood each other, and I was comfortable with him. And attracted: his eyes were the color of the sea and his blond hair and tan skin were the color of sand. Just like Israel, his birthplace. He wasn't big, maybe 5'7" and a lean 125 pounds. But he was well built, proportional, elegant. A *sexy* Israelite. Though he was a sophomore and I was a junior, he was two years older than I, on account of Israel's conscripted military service after high school.

But was the guy gay or straight or bi- or tri- or uni- or poly-

or a-, or what? He did seem to bask in the unrequitedness of Miles's ardor, and had no girlfriend. So was Miles my competition? And if so, how could I outdo gay love? Was I gonna have to fight a sodomite for my Israelite?

"*Atjar bening,* honeys," Miles said, placing a plastic bowl of what appeared to be pickled vegetables before us. "Delish. Tart and sweet."

The Israelite and I looked at each other and tentatively tried a cucumber slice. Then we both lunged for our drinks. Then we started laughing.

"What am I missing, what am I missing?" Miles asked. Explaining each dish in detail, he excitedly set down bowls of *bami goreng* (fried egg noodles), *sambal goreng telor* (a hard-boiled egg dish with coconut milk and tamarind paste), and *boeboer ajam* (savory rice pudding with shredded chicken). One entrée was more insipid and awful than the next. It all looked like gruel. Everything was wrong with everything: underspiced, overspiced, watery, dry, undercooked, overcooked, no harmony or balance among ingredients. I'm sure Indonesian food is fine when properly prepared—most foods are. But this was not one of those nights or one of those chefs. By the looks of it, the Israelite concurred with my negative assessment.

"You two angels are eating like BIRDS," Miles said, helping himself to more gloppy *bami goreng.* "Are you saving room for dessert?"

"I know I am," I said, wiping my mouth with a paper towel.

"Good," Miles said. "Because I made cheesecake. Lemon-lime!"

The Israelite excused himself and went to the bathroom while I helped Miles clear the table and urgently whisper-gossip in the kitchen.

"Is the food really that horrible?" Miles said.

"Yes."

"But isn't the Israelite good enough to eat?"

"Yes," I said. "Are you sleeping with him?"

"Not yet. I'm trying to reach his heart through his stomach."

"That won't work," I said.

"You're just jealous," Miles said, clearing his throat to signify the Israelite's reemergence.

"Anything I can do to help?" the Israelite said.

"No, darling," Miles said. "Sit that beautiful Middle Eastern Sabra ass down and relax. We're getting you coffee and dessert."

The lemon-lime cheesecake was runny, room temperature, and salty. Miles devoured his slice and helped himself to another as the Israelite and I dragged our forks through the sickly muck, and drank coffee and smoked cigarettes. Miles began a curious game with the Israelite. Whoever lost had to pay the winner a dollar. It was like watching a cinematic Ping-Pong match. Miles served first.

"Renoir or Bergman?" he said. "And no Americans this time, sweetie."

"Buñuel," the Israelite said. "Bertolucci or Truffaut?"

"Godard," Miles said. "Visconti or De Sica?"

"Cocteau. Eisenstein or Fellini?"

"Anderson. Lean or Wenders?"

"Kurosawa," the Israelite said. "Herzog or Lang?"

"Polanski. Wertmüller or Rohmer?"

"Ophüls. Lelouch or Antonioni?"

"Chabrol," Miles said. "Angelopoulos or Kubrick?"

"Kubrick's an American," the Israelite said. "Technically."

"He is?" I said. "I thought he was British."

"British by way of the Bronx," the Israelite said. "One dollar, please."

"I should never mix beer and wine," Miles said.

And I should never mix gay men with sexually ambiguous ones. Either way, I was ready to smack 'em both upside the head. It's like being with people who are speaking to each other in a language you don't know. It's insufferable. Still, it was interesting to hear about these foreign directors. My un-American movie knowledge could use some furthering, as I was exclusively studying English and Art History.

"I liked *Tootsie*," I said. "Who directed that?"

"Sydney Pollack," they chimed. "But he's an American."

"How about *Diner*?" I asked. "I loved that one."

"Barry Levinson," they answered. "Also an American."

"Yeah? Well you know what?" I said. "You guys are doing something from *Diner* and you don't even know it. Eddie and the Colts?"

"You mean, like, B-B-B-Bennie and the Jets?" Miles said, looking perplexed.

"WHAT are you talking about?" I said. "I'm talking about *Diner*, not a freakin' Elton John song!"

"Okay, we give up," the Israelite said.

"Eddie in the movie? Steve Guttenberg? He's engaged to this girl named Elyse. And he's such a Baltimore Colts fanatic that he insists that their colors be the colors for their wedding."

"What are the colors?" Miles asked.

"That's not the point," I said. "Okay, so Steve Guttenberg is so panicky about getting married that he makes Elyse pass a sports test. I think she's gotta score, like, at least a sixty-five. And if she flunks, he won't marry her."

"Okay, honey, okay," Miles said, still utterly perplexed. "Don't get upset."

"Hmmm," the Israelite said to me. "Your rabble-rousing mind works in very mysterious ways. Interesting. Different."

"Betcha neither of you could do this for Art History," I said. "Although you could start with Renoir in my field. *I* wouldn't. But you could. Renoir's for babies and old ladies, that's what my Art History professor thinks. Now Manet, there's a real artiste."

"Your Renoir was the son of our Renoir," Miles said, standing up to dig for change inside his jeans pocket. "Renoir *fils*."

"Duh," I said.

"Wouldn't you rather have a kiss than money, honey?" Miles asked the Israelite, batting his purple-tinted eyelashes like a tarty Elsie the Borden cow.

"Gotta call it a night," the Israelite said, rising. "Got a lot of studying to do."

"Don't forget, love," Miles told the Israelite, caressing his cheek. "We're seeing *The Marriage of Maria Braun* on Tuesday night."

"I know," the Israelite replied, pulling on a burgundy and gold Redskins cap and a faded blue denim jacket. "Fassbinder."

"*I feel* like we're hiding from Miles," I told the Israelite, hanging up some summer dresses in his closet. "It's like we're sneaking around behind daddy's back."

"We are," the Israelite replied. "Hey, I'm making room for your three thousand additional feminine products. It's gettin' tight in this little bathroom. But we'll manage. What are these instruments of torture?" He held a glass containing a Shu Uemura eyelash curler and a pair of pink Tweezerman tweezers.

"What about Miles?" I asked.

"Miles is in love with me, or at least infatuated. He knows I have no interest in going that way. It's almost sad. It's a lost cause but he can't help himself. It's agony for him."

"Well, why don't you talk to him about it?"

"We don't need to discuss it. He's always known."

"How?"

"From my talk about girls. And he doesn't want to talk about it. He doesn't want to ruin it."

Neither did I, actually. For two such talky Jews, there was one area the Israelite and I never touched on, in spite of our increasing intimacy: the fact that we weren't having sex and how rejected that made me feel. Since that first night at Miles's, the Israelite and I had become inseparable. I'd gradually begun leaving a few hundred of my critical personal items—toiletries, cosmetics, TaBs—in his Takoma Park apartment. It was easier than having to schlep them over every time I stayed overnight, which was often. I loved being at the Israelite's place. It wasn't fancy but it was, as Hemingway would say, clean and well lit, and obviously homier and more private than those scary student dorms or off-campus housing. The Israelite and I, I and the Israelite. It had all seemed so natural, falling in together. After our leaving Miles's house in separate cars a month ago, the Israelite had honked at me to pull over. He wanted to know if I was hungry. I was. We drove to Ledo's and gorged ourselves on toasted meat and cheese raviolis with "robusto marinara" dipping sauce, and that glorious pizza. It was so good it made me wanna holla an aria. And when I got cold—I always shiver after I eat, sometimes to the point of such teeth-chattering, frozen nose- and finger-tipped anguish that strangers think I'm seizing and kindly offer to call 911— the Israelite put his faded Levi's blue jean jacket around my shoulders. He noted I might be a tad inappropriately dressed for the weather.

"What do you mean?" I asked. "It's March. It's spring, dear."

"Yeah, but you're in a light dress without sleeves. I'd be cold."

"Yeah. I probably should've brought along a jacket or a little sweater or something. But it was so beautiful out today, it was warm. And I love this dress!"

I really did. It was a great dress to (very carefully) eat a lot of pizza in, a full sleeveless swing dress with an empire waist in white cotton floral lace with thick, wide straps and a ruffled tiered hem hitting about mid-calf. It was an artisanally groovy, calorically forgiving work of art. Granted, it was more sincerely summery and less transitional than strictly springy, but there are times when you just can't wait any longer for the next season to come. I wore it with these really pretty teardrop pearl earrings, vintage Mexican inlaid abalone and silver cuffs, super-cute black leather wedge sandals, and tomato red-painted toenails. The Israelite's denim jacket actually went perfectly with my outfit. The fact that it did made me think, *God whispers before God shouts. This is DES-TINY. When everything coordinates effortlessly, that isn't mere style. That's LOVE. BIG LOVE.*

So I'm shacked up with a beautiful Israelite and he won't make love to me. How insanely frustrating is that? Because you know this is not normal. Israelites love sex. The men are animals! When God said to be fruitful and multiply, they went for it. Even the Reform ones. So what was wrong with mine? I was doing the Lucy Ricardo whine all the time—*waaah.* Seriously. We lived together, we used the same tiny bathroom, we did housework to-gether (he loved cleaning, can you stand it?), we shared our matzo (excellent when heated in the oven for five minutes and crumbled into tomato soup, a five-star repast compared to anything from Miles's kitchen), we both loved the Pretenders and the Police and the Redskins, we *slept in the same bed.*

No sex.

I know what you're thinking. It's what I was thinking. But gay would be too easy—and *nothing* in my life is easy. The real problem, it seemed to me, wasn't my Israelite's sexual orientation. It was my Israelite's Israeli ex-girlfriend, his first serious love. The *femme fatale* specialized in prey traumatization. Remember that Mademoiselle X chick that Miles had told me about? That's who this was. My Israelite had met her right after high school, when they both served in the Israeli Defense Forces. He was a flight mechanic working on aircraft like those cool Cobra helicopter gunships. Mademoiselle X was a secretary in the same squadron. She never saw combat but she knew how to shoot an Uzi and not miss. She was also completely out of her mind *and* a Nastassja Kinski look-alike. This combination was not helpful to my cause. Neither was my beloved's fascination for the real Nastassja Kinski. In the apartment's entryway, he'd hung a long horizontal framed poster of "Nastassja Kinski and the Serpent," the famous 1981 Richard Avedon photograph of a prone, nude (except for a chunky ivory bracelet) Nastassja with an enormous python snaked around her body, its tongue flicking her ear.

Welcome to my world.

Bottom line: Mademoiselle X had really messed up my Israelite. That's what *he* blamed his sexual problem on, anyhow. Though she was two years younger than he, Mademoiselle X was far more experienced in the ways of the world and had taken the lead. She was a take-charge kind of man-eater (which I guess is kind of redundant, Hall & Oates songs notwithstanding). My relatively innocent Israelite didn't stand a chance. A year and a half into the relationship, Mademoiselle X brusquely left him and traded up for a pilot, *la crème de la crème* of Israeli society. When that affair crashed and burned three months later,

Mademoiselle X scurried back to the lowly mechanic. It was a use-less scurry, though; there was nothing there. The Israelite was a broken lowly mechanic. *Un mechanique tragique!* The mechanic couldn't even figure out how to put himself back together. He preferred keeping the black-and-white photographs he'd taken of Mademoiselle X, to keeping Mademoiselle X. I know, because he'd shown me the moody pics. As if being photographed in film noirish tones, wearing black clothes, and having a well-defined clavicle makes you deep!

I'm sorry, but some women are romantic slobs. It's true. They're careless. They expect other people to pick up after them and clean up the broken messes they leave behind. This one had turned my Israelite into Humpty Dumpty, and he'd had a great fall. *All the King's horses and all the King's men couldn't put* le mecha-nique tragique *back together again.* But I thought I could. I had Krazy Glue. I could Krazy Glue those pieces back together. Thanks to one Mademoiselle X, my Israelite had gone from utter confusion to zero confidence to sexual paralysis. That's what lucky gals like *moi* inherit. I either had to live with that mess or clean him up.

I'm into the cleaning mode. I *love* to clean. I'm really good at it. Maybe I'm a latent housewife. For example, I actually took plea-sure recently in buying a new vacuum cleaner at Costco (Eureka Boss SmartVac with the Power Paw Turbo Brush—fabulous). I also was really excited to discover Windex in the "lavanda" scent, which is purple and smells just like lavender with a New York ac-cent. I had to think: What beyond carpet and floor-care appliances and aromatherapy cleaning products would de-Mademoiselle X-ify my Israelite? It would have to be good, very good, maybe even a little kinky, like Kinski with the serpent. But no reptiles. Please. They're disgusting.

Gross-out factors aside, that might be what it would take. After all, the Israelite did trace Nastassja's serpent's tail, tucked neatly between her shapely thighs, whenever he left and entered the house. It was like his own personal mezuzah. I called one of my best friends, the Morbidly Mystical Mexicana. Where she's from, they know their herpetology. Maybe there was some exotic south-of-the-border move I didn't know about, some secret snake charm.

"Your Israelite can never be free," she said. "He's *una alma perdida.*" A lost soul.

"Okay, he's the Jewish Travis Bickle, king of Hebrew pain, whatever," I said. "Is there some kind of awesome Mexican sex thing we can do that I haven't covered? Something you crazy kids do down there with the snakes and lizards and iguanas? Like, *Night of the Iguana?*"

"Um, iguanas *are* lizards," La Mexicana said.

"And you guys eat them, right? Like, does it give you a special power or something? A love potion *número nueve?*"

"I don't think you can fix your iguana," she said. "I'm sorry, honey. I'm worried for you, not for him. I'm worried for your heart. How is your heart? Are you taking care of it?"

I took a long drag off my Parliament and exhaled.

"Dear, is there some psycho folk healer down there you know?" I asked, reaching for my TaB. "Some powder? Some pill? Some prayer?"

"*M'ija*, you've lived with him since MARCH."

"I know, beware the Ides of March. Is that where you're going with this? That I'm gonna be assassinated like Julius Caesar and you're, like, the soothsayer-dreamer guy whose prophecy I disregard to my own debauchment?"

"Okay, I have no idea what what you just said means."

"Sorry. *Julius Caesar.* We're reading him in my Shakespeare seminar."

"Oh. No, what I was saying is that it's summer. That's *months* of no sex, honey-bunny, *months.*"

"We have sex," I said. "Just not with each other. I mean, you know, sex *alone.* I would've jumped out the window otherwise."

"Why don't you talk to him about it?"

"I can't talk to him about it."

"Why not?"

"It would ruin everything. It's too personal."

"Chica, it sounds like it's too late for *curanderos."*

I love my Mexicana but we don't call her Morbidly Mystical for nothing. She believes in *curanderos,* folk healers, and she *loves* death. Every day's *Día de los Muertos* for her. It's really shocking she isn't more into Shakespeare. It's the bloodiest stuff. Look at Cleopatra and the asp. *Give me my robe, put on my crown; I have immortal longings in me.* Such poetic words. Then she puts an asp on her breast and dies from its bite. La Mexicana would love that drama, she'd be in heaven.

Hey. Wait a minute. Cleopatra-asp. Nastassja-serpent. Helicopter-Cobra. If I could just find something serpentine to wear, maybe that would do it! Or even subliminally serpentine, like in the Hebrew bible, where snakes are subtle and illicit, but not Satanic. *I've got the apple of temptation,* Joni Mitchell sang, *and a diamond snake around my arm.* I'd seen a picture somewhere of my musical heroine wearing just such an armlet. I'd seen slides of snake jewelry like Joni's in one of my Art History classes. The ancient period. There are reliefs and paintings on the walls of royal and high-ranking Egyptian tombs that illustrate how jewelry was made and worn. The beautiful diadems, earrings, necklaces, pendants, armlets, bracelets, anklets, and hair ornaments not only dolled you up, my brilliant professor explained, but were considered amulets, protecting you from harm.

"Haven't serpents been known to eat their tails?" I asked.

"The semiprecocious Gigi Anders has spoken," he told the class, smiling and clicking the slide projector in the dark. Notice how he didn't answer my question, nor did he interdisciplinarily mention *Antony and Cleopatra*, and that's *all* about Egypt.

I drew a series of slithery figure eights in the margin of my notebook. A snake-print dress? Some snake-print jeans? A pair of sunglasses with snake temples?

Who was I kidding? I get creeped out just walking past the Reptiles aisle at PetSmart. Damn. Some men require you to re-think your initial fashion impulses. My man, with his shy mystery and elusive Eros, was one. I knew I needed a different design altogether to penetrate that aloof aura and make my Old Testament Israelite love me biblically. Something more . . . I don't know, warm-blooded.

"*How often* does he get high?" asked my therapist, Manny. We were discussing my sex life. Or rather, lack of.

"Every day," I said. "Like, after dinner. Sometimes in the af-ternoon, too. He gets his pipe out and smokes hash. Hash browns, I call it. I tried it once. So harsh. I hated it."

"Every night before bed, then," he said, "he's stoned."

"Yep. For hours. He zones out. Sits there on the sofa, all se-dated, drinking Turkish coffee and smoking Kents and listening to the Pretenders with the TV on without the volume. And I mean, I like the Pretenders. That Chrissie Hynde is really—"

"You know that all that weed can make some men impotent," Manny said.

"No! Really? So that's what it is? It's not Mademoiselle X?"

"If he stops smoking hash browns he can get his erections back almost immediately," Manny said. "That's the good news. As for the Mademoiselle X-planation, I don't buy it."

"You DON'T?"

"Nope. Smells fishy to me."

"So what are you saying? You think he's GAY?"

"I don't know if he's gay or a voyeur or androgynous," he said. "I've never met him. But I don't believe his stated reason for not being with you romantically. If he can masturbate to a climax—"

"He claims he smokes hash browns to escape, that he feels inept, like a failure," I said. "He keeps reading Hermann Hesse, saying he identifies with his gloomy characters—yikes. And he drags me off to see all these arty sex movies like *Last Tango in Paris*. Do I have to love that movie just because Pauline Kael and my boyfriend loved it? He's so down on himself. He wants to be a filmmaker. But then he goes, 'The artist is like a little dictator and I don't know that I have it in me.' He's so defeated. It's really depressing. And yet he likes to shoot me all the time."

"Interesting choice of words. Shoot you? Shoot you naked? Shoot you down?"

"Well, sometimes naked. Not always, though. It depends. He's like this beautiful, intelligent puzzle. I'm his muse, that's what he says. He recently framed and hung a picture he took of me. I am dressed in it, by the way."

"Dressed in what?" Manny asked.

"A white dress," I said. "It's a white lace dress. I was wearing it the night we first met."

"He may *picture* you as a bride. But only in pictures. Not in reality."

Oh. Shit. Once again, Manny was right on the money. I hate that.

There's a wonderful photograph I've loved for years. It's quite famous, by Robert Capa. Shot in black-and-white on a brilliantly sunny day in 1948, Pablo Picasso, his beautiful forty-years-younger French lover, Françoise Gilot, and Picasso's nephew, Javier Vilaro, are walking away from the water on the beach in Golfe-Juan, France. Approaching the viewer, the three figures are staggered across the composition from left to right, biggest to smallest; first Françoise in the foreground, then Picasso, then Javier down by the strand. In an open short-sleeved shirt and knee-length shorts, the barefoot Picasso is holding a huge fringed beach umbrella aloft, shading Françoise, holding it as you would do for Aphrodite, for Cleopatra, for Deneuve circa *Belle de jour.*

The iconic picture's always captioned "Picasso with . . ." or "Picasso and . . ." but clearly Françoise is the main event here. She looks so happy, regal as an empress and fresh as lavender. She's wearing a big straw hat with a frayed, deliberately unfinished brim; something ornamental around her neck—could be a twisted scarf or a chunky white bead necklace or a string of bleached seashells, hard to tell for sure; and a pale, cap-sleeved, fitted, ankle-length cotton dress.

Perfection.

"It was a time of contentment," she once wrote. Indeed, it was Françoise's book, *Life with Picasso,* as well as that Capa portrait, as well as the fact that I was immersed in Art History, that had inspired me to become the ideal latter-day Françoise to the Israelite. I'd be the best muse anybody ever saw. I'd just have to figure out how to move us to the next a*mus*ing level. But unlike

Françoise and Picasso's, I wanted ours to be a happy ending. Because of all Picasso's many mistresses and wives, only Françoise, whom he never married, left *him*. He was pissed, too. He never forgave her for it. Françoise said that after a decade it was no longer fun living with a monument. By then Picasso had become "a sacred monster." Reproductions of his drawings and paintings and *objets* appeared on every museum gift shop's coffee mugs and wall calendars and coaster sets. The yuppification of *l'artiste*. What was once pure had become stained. Françoise ran away with their two little kids, Claude and Paloma, and eventually married Jonas Salk. Don't you love it? And don't you love Paloma? She's fabulous. Paloma has her own eponymous perfume and she designs jewelry for Tiffany. She loves red lipstick so much she invented her own and called it Mon Rouge. My Red. Years later, I interviewed her for the *Washington Post* Style section for a long story I wrote about the history of red lips. Paloma was so nice!

I opened my Capa book and stared at Paloma's beautiful mother on the beach. In my imagination, Françoise's pretty summer dress was backless. That was her warm-blooded secret. Picasso was actually staring at Françoise's bare, tanned back. That's why the sacred monster looked so uncharacteristically happy. *The artist is like a little dictator and I don't know that I have it in me.* Yes you do, dear. You just need the right stimulation. And trust me, it won't involve snakes.

Theory. Theory came to my rescue. Also to my near bankruptcy. I was on my break at Bloomingdale's in White Flint, Maryland. I worked at their Clinique counter part-time. I hated the white polyester lab coat I was required to wear, but I loved my job. *Beaucoup* free samples, that handy 20 percent employee discount, and I was the most in-demand makeup artist on the floor. A lot of cosmetics salespeople have absolutely no idea what they're doing. If you tell them, "I need a warm red lipstick," they think you mean you need it microwaved. *Clueless* about color. I'd always tell my customers, "Don't buy something from someone whose makeup looks like RuPaul's." I'd hear the word *neutral* a lot, as in "M·A·C's Spice is the perfect neutral lip liner." No it's not. Live lips are not mauve-brown. There's no such thing as a "neutral" anything when it comes to color (except for periwinkle blue, which I wouldn't recommend for lips). This is why women get so confused. They get bombarded by fashion magazines and celebrities on the Internet and on E! and see all this stuff in Sephora or on their girlfriends—and randomly buy it, thinking it's right for them. And then they wonder why everything they have doesn't work and they run to get Botox and Restylane and face-lifts. (Not that I'm against plastic surgery. I think it's fantastic, as long as you can't tell it's been done and your practitioner is Roger Oldham, *my* practitioner. Dr. Oldham RULES.)

So I'm wandering through Bloomies's day dress department and, as Blanche DuBois would say, "Sometimes—there's God—so quickly!" It was heaven-sent. It was divine. It was the most awesome Little Black Dress (LBD) you've ever seen. I mean raise-the-Israelite-from-the-dead-and-create-a-dictator *amazing.* (And not a snake in sight. Theory's not a snaky kind of brand, thank you.) This was the ultimate addition to my already considerable feminine arsenal, in viscose with a touch of spandex. Jewel neckline. Short sleeves. Dropped waist, with a shirred band of the same fabric wrapped around the hips. A flouncy, just-above-the-knee hemline. So cute! But wait, it gets better. This is the killer feature: turn it around and this LBD was . . . BACKLESS. Just like I'd imagined Françoise Gilot's was in the Capa photograph! This wide-open back was geometric, almost architectural; a crisply cut-out vertical rectangle across the entire back down to the small of the back, like a very low cut bathing suit back. (It had a 2"-wide horizontal band across the shoulders on top keeping the sleeves together.) Like a Rudi Gernreich peekaboo dress with an air of Yves St. Laurent during his Mondrian phase. DEVASTATING. Daring from the back, demure from the front. I looked at myself from different angles over my shoulder in the three-way dressing room mirror. Take that, Mademoiselle X. Take that, Nastassja. Take that, Miles. Oh God, poor Miles. This dress would be the death of him. I felt so guilty about stealing his boyfriend. But, *c'est la guerre.*

The Israelite and I had planned to spend the following Saturday in Annapolis, as neither of us had ever been there and we both loved long drives. Hmmm. Backlessness at the Maryland state capital? Could one Israelite and a brigade of 4,000 midshipmen from the U.S. Naval Academy handle it? I was so high over my little black Theory dress that I didn't even look at its big white price tag. I foolishly figured that with my employee discount I could swing it. I handed the saleslady my credit card, squeezed my eyes shut, and sent up a prayer for mercy. Apparently, God was out on a break.

"Okay. $327.60," the lady said. "It's really cute. I love this dress."

"I'm sorry," I said. "WHAT did you just say?"

"That it's adorable?"

"No. The other thing. The thing with the numbers in it."

"Oh. $327.60," she said, opening a Medium Brown Bag.

"Did you do my discount?" I said. I was starting to sweat.

"Yeah."

"But—"

"Are you okay?" she said, touching my arm.

"How can it be that much?!?" I screamed. "That's impossible! It's missing a back! How can I be paying that much money for something that's not there?!?"

"Well, the dress is $390—"

"It IS?"

"Minus your 20 percent, that's $312. Plus 5 percent sales tax is $15.60. Totaling—"

"DON'T SAY IT AGAIN!" I screamed. I tried to mentally calculate how many Dramatically Different Moisturizing Lotions I'd have to sell to pay for my dress. Let's see, at $21.50 a bottle, it would take . . . oh, the hell with it. Give me my Theory, put on my lipstick; I have immortal longings in me. Call it the Big Bang Theory.

I saw sailboats in his eyes. Dinghies and schooners, too. Well, I saw their reflections on the lenses of his American Optical aviator sunglasses. The Israelite and I were having a crab cake lunch on bustling Compromise Street, just south of the city dock, right on the historic Annapolis waterfront. *Compromise* Street. Yep, that was the story of our relationship, all right. Though it was balmy out, a perfect summer day, I'd put on the Israelite's jean jacket (with the cuffs rolled back, of course) over my LBD before we left the house. I wanted to wait for the exact right moment to flash him, as it were, into virility. I'd decided to nix a bra—it would ruin the Theoretical line—and went with a pair of chandelier earrings that resembled peacocks' spread fans, big black sunglasses, very red lips and toenails, and a fanciful pair of L'Autre Chose black slingback espadrilles with 4" black jute wedge heels, platform soles, black leather bands across the open toes, and decorative little black pom-poms down the front. A marine motif, in other words.

I finished my raspberry iced tea, got up, took off the jacket, walked to the railing overlooking the tall white ships in the blue Chesapeake bay beneath the blue Maryland sky, and leaned on it for maximum exposure. I heard the Israelite flick his Bic.

"That's a very unusual dress," he said, after a little pause.

"Yeah?" I said, not turning around. I felt his eyes and the sun on my back. It was great. Like most inner and outer feelings, this one reminded me of Leonard Cohen lyrics: *I love to see you naked over there, especially from the back.*

"You're exposed," the Israelite said. "You're exposing your back story." I heard him unzip a compartment in his backpack. Like a trained model, I kept my pose. He snapped a few pictures and put the camera down.

"Should I turn around?" I asked.

"That's your prerogative," he said. "I ain't gonna get between you and that dress."

"Why not?" I asked.

"I've seen you wearing unusual apparati before," he said, sipping coffee. "So I'm only mildly shocked. You like making slightly inappropriate fashion statements. You're outrageous, it's who you are."

I wasn't sure how to take that. Unless a man says, "You look stunning, you're the envy of every woman here, I'm so proud to be with you," I assume he's being critical. I stared out across the water at the seagulls. I felt as I have in those dreams where I'm the only undressed one in public. Two uniformed Naval Academy midshipmen who'd been sitting nearby approached the Israelite and told him in no uncertain terms that "that" was "unacceptable."

"Look," he told them, "you might be naval types but I'm an air force type. I see that you're unranked. I'm a veteran sergeant first class. So before you want to meet the same end as the others, fuck off."

They saluted and left.

"My hero!" I cried, rushing over to hug my Israelite.

"Have some pie, honey pie," he said, passing a plate with a big wedge of Key Lime on it. "It's much better than Miles's lemon-lime fiasco." He unzipped another compartment in his backpack, produced a small midnight blue velvet jewelry box, and handed it to me. I put my dessert fork down. "A little token of my appreciation," he said. The box make a soft *crrrack* sound as I slowly opened it. It was a diamond ring. Pink gold filigree in the shape of a peony. Its bud was a diamond.

"Oh my God!" I cried. My Theory! It was working! Overtime! I KNEW the snakes would be all wrong!

"You like peonies," the Israelite said.

"I love peonies," I said, my eyes watering. "I love you."

"Good. Try it on."

"Wait. Is this a . . . I mean, are you . . . ? "

"You want to get married?"

"Um, well, uh, yeah," I stammered. "Theoretically."

"You sure?"

"It's a really beautiful ring. Really beautiful. Unique. I'm just . . . ah."

"What?" the Israelite said.

"I, I'm just gonna say this, okay? I can't marry a man who's not my lover. I just . . . I love you. But I just can't."

"You like those ships out there," he said unflappably. "I thought we could sail around the world for a long honeymoon. I inherited a little family money a while ago. I could pay off all your credit cards."

He had me there. Plastics. The scourge of my very existence. I suddenly realized my back felt hot. Oh no. Oh nooo. Oh fuuuck.

"I think I need Solarcaine," I said, turning to show the Israelite my back. "With the aloe. Look. Is it bad?"

"It's not good," the Israelite said. "Does it hurt?"

I started to cover myself with his jacket but the fabric against my sunburned skin made it worse. *To burn with desire and keep quiet about it is the greatest punishment we can bring on ourselves . . .*

"I can't take this," I said. I meant the ring and I meant a lot of things. Do serpents really eat their tails? I don't know. But they're molters, I know that much. They shed their skins. It was time for me to, too.

I put my LBD back in the closet. Like my Israelite.

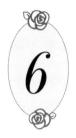

6

COCO

Do you feel dressed without perfume? I mean well dressed? I don't even feel well *un*dressed without it. Perfume: The fragrant period at the end of the feminine sentence. Or maybe, since it's an invisible, potent accessory that trails along the senses, it's an ellipsis. . . .

Either way, I need it to feel finished. Even if no one's around to smell me. But someone's always around. Sometimes that's the problem. Three people (none of them lovers) have alleged they're "allergic" to my perfume. A fourth said it was "weird" that she could smell my perfume in my hair. Obviously, they're all dopes. They don't deserve me or my COCO, my favorite and the best perfume of all time. The perfume so confident it comes in all caps.

B.C. (Before COCO), I was a serial monogamist, with the sporadic slip into perfume sample sluttiness. This is appropriate, considering I usually lean toward French perfumes. Unlike Lerner and Loewe's Gigi, I *do* understand the Parisians, *Making love every time they get the chance, Wasting every lovely night on ro-*

mance. (Note how the lyrics don't say *to* or *with* the same person. Nor do they point out that Chance and Romance are both the names of nice perfumes by Chanel and Ralph Lauren, respectively.) In junior high and high school, while the other girls were into white lipstick and Yardley of London's Love, Revlon's Charlie, and Jōvan's Musk, I was *enchantée* by screamingly red lipstick and Yves St. Laurent's Rive Gauche. Its ad campaign had a beautiful young French coquette spritzing it on and saying, "I'll be . . . me." I'll be *moi*. My *moi* was vibrant, modern, feminine, silvery-cool, light, fresh, oak mossy roses, and myrrh. Loved it. And loved *moi*'s container, a cylindrical metal tube in bright turquoise ringed with a descending trio of black stripes and bordered top and bottom with silver stripes. So chic! Chic perfume bottles are key. They're part of the whole allure of fragrance, your womanly mystique right there looking pretty and mystique-y on your boudoir dresser or in the bathroom. They're like jewelry for countertops.

One day I decided to cut French class to go loiter in Saks's cosmetics department. I had this older French friend who worked there named Élodie. I'd go visit her whenever I felt the need for a Gallic field trip, as it were. Élodie was this gorgeous, tall, slim blonde who worked at the perfume counter. She didn't really need to work at all; her married lover paid for everything, including her *fabuleux* Georgetown apartment. A kept woman! *Une femme gardée!* I'd only ever read about such exotic creatures in *Cosmo.* Élodie was very glamorous and sophisticated, very world-weary and worldly and red-lipsticky. I considered her my true mentor in all things French, which meant I wasn't *really* cutting French class when I fled school to spend time with her. No guilt. Guilt is *so* un-French. You saw *The Sorrow and the Pity.*

Élodie had a porcelain complexion and amazing gray-blue

eyes whose inner and outer lids were rimmed in blackest kohl. As the day wore on, her huge eyes got softer, smudgier, more enigmatically *I-just-had-great-sex-and-a-bottle-of-Veuve-with-my-subversive-lover-who-looks-like-Olivier Martinez* seductive. This of course only works on Frenchwomen. If I did that kohl like that I'd look like a demented raccoon, a common prostitute, or some other small nocturnal omnivore. Élodie liked getting out during the day and having a boyfriend-sanctioned place to wear the boy-friend-procured Chanel couture (i.e., it was okay to look like a runway model as long as she never met other men while she looked like a runway model). Élodie also liked earning a little pocket money of her own—"Just in case, one never knows in life"—and she definitely knew her perfumes. Being with Élodie was like learning about wine from an oenophile. She taught me the differ-ence between perfumes' three primary "notes"—top, heart, and base—and perfumes' four "families," or basic categories based on ingredients (and there are a million subcategories): fresh (citrus, green leaves, marine-y), floral (mixed soft bouquets), oriental (spices, vanilla, patchouli), and *Chypre*, or woody (honey, earth, tobacco).

Élodie understood taste, personality, style, nuance, when to splurge and how and on what. Knowing how to wear good per-fume and choosing the right one was like knowing how to tie an Hermès scarf a hundred different ways—*toujours* a must, but it has to be the right one for you. She could wade into vast seas of fra-grant possibilities in no time and resurface with an edited, select, perfect one or two. Once, I asked her if I should just buy Chanel N°5 and get it over with. You know, stop catting around on Rive Gauche with all the distracting samples and just commit to N°5 for life. Simplify. Grow up. Exemplify the poster of the Andy War-hol silk screen print of that iconic perfume I had hanging on my bathroom wall. After all, Marilyn Monroe, another Warhol idol, said a few drops of Nº5 was all she ever wore to bed. But Élodie frowned and shook her head. She said that Chanel's *Numéro Cinq* was indeed a classic, but it was the sort of classic you'd give to your mother for Mother's Day. To wit, it was wrong for *moi*.

"I want for you to smell this, *mon petit chou* [my little cabbage]," Élodie said in her charming French accent. Listening to her talk was like listening to charms' tinkling crashes on a woman's bracelet. It was Chanel's new light moss green-colored scent, simply called Nº19. She sprayed the *eau de parfum* on my inner wrists.

"Mmm!" I said. "*Il sent merveilleux!*" It smells marvelous! Since I was taking French in school, Élodie was great to practice on. "And I love the flacon. So sleek." It was a vertical crystal rectangle with a brushed silver cap, and that simple, modern, elegant black block print that appears on all Chanel products.

"*Pas mal, eh?*" Élodie said. Not bad, huh? The French are impressed by *rien*. Nothing. Perhaps they consider overt enthusiasm jejune, an exceptionally un-French trait. In Élodie's presence I often felt like a Henry James heroine, personifying the archetypal collision of American innocence with European experience. *Pas mal. Pas mal* is the French equivalent of an American multiple orgasm. "This one was created for Coco Chanel," Élodie continued. "It was her personal fragrance. Isn't it witty? Unconventional. It's you. *C'est toi.* A little audacious. Here, try a bit more. Let it rest a moment. Savor it."

"Mmm!" I said, bringing my wrist to my nose. "And why is it '19'?"

"For Coco's birthday of August nineteenth. A real Leo, *n'est-ce-pas?*"

"It's different than Rive Gauche," I said. "More . . . something. It's delicate. And youthful. But it's self-possessed, too. Is it a floral-woody-green?"

"*Voilà*," Élodie said, nodding approvingly at her perfume pupil. A few strands of her long straight bangs strayed into her long black eyelashes. As she blinked, the strands blinked along, too, but stayed there. I wish I had straight blond hair and black-rimmed eyes. "It is woody, a trace. And it has the jasmine, rose, iris, ylang-ylang, sandalwood, and mosses."

I found all of this aromatic lore infinitely more interesting than perfecting my *passé composé* and *imparfait* in a junior high school classroom.

"Your man doesn't deserve you, Élodie," I said. "Neither does this store. You should be working at French *Vogue* or with Jacques Polges on Rue Cambon and having an affair with *him*." Jacques Polges, Élodie had once explained, was the "nose" of the Chanel design house, located in Paris on Rue Cambon.

"You will take these home, *chèrie*," Élodie said, wrapping several tiny N°19 glass vial samples in tissue paper. She always gave me tons of samples, always French ones. It's like she *wanted* me to be promiscuous. You had to respect that. I had vials and vials of Robert Piquet's Fracas, Guerlain's Shalimar, Jean Patou's Joy and 1000, Houbigant's Aperçu and Quelques Fleurs, and Hermès' Calèche. I also had a petite legion of Annick Goutal (Ce Soir ou Jamais and Quel Amour! and Hadrien Absolu, which is technically for *les hommes* but we don't care because it's exactly like Eau d'Hadrien Femme except it comes in the handy, cost-saving, longer-lasting Big Bottle), and Christian Dior (Diorissimo, Diorella, Dioressence). They were all fabulous variations on the French theme of *fabuleux*, especially Hadrien Absolu. (That's what Élodie wore, though she layered it with others. Like her home address and the precise nature and details of her love affair, she kept her formula a secret. That's so very French. They love culti- vating that aloof *mystère*.)

But for my nose none beat this new Chanel N°19, not even my trusty Rive Gauche. *Pardonnez moi*, Yves, but N°19 was a revela- tion. It was Rive Gauche matured and out of high school, with a better wardrobe, higher heels, and a more interesting beau. "Take it home and try it out," Élodie said with a shrug. "You play, you sense, you see."

That's the French way.

"No, I'll buy that 19 right now, dear," I said, digging in my purse for my EVC, or Emergency Visa Card. "I love it. I can tell it's the best one. Just don't tell me what it costs. I don't want to know."

That's the American way.

Thirteen years later, in 1984, the biggest news rocking my world wasn't Madonna's "Like a Virgin" or Ronald Reagan's reelection or Apple introducing its Macintosh personal computer. Nope, nope, and nope. It was COCO. Chanel's dreamy new perfume, also created by Jacques Polges, was cognac golden, baroque, and expensive. Its bouquet was totally different from N°19's. More lingering. Deeper. Fuller. Saturated. Rather than mossy green woods and flowers and summer sunbeams dappling through leafy treetops and sweet little stolen love kisses in secret places, it was a languorous, womanly purr. It was purple velvet luxurious and winter nightfalls in the city and red wine- stained lips. It was like a gorgeous, erotic fever, that crazy desire the French call *amour fou*, flooding you. *Après* N°19, *le* COCO *déluge*.

I was first introduced to COCO not by Élodie but by *Vogue*. There was an ad in it with a scent strip. I usually tear these things out, sniff them briefly, and throw them away before I can actually sit down and enjoy reading the magazine. Not just the scent strips, either. I'm compulsive about thinning out all advertisements from magazines. Ads make me insane, especially those thick, cardboardy subscription cards. They either go into free fall all over the floor, or else completely antagonize me by clinging to the magazine's spine. Ads with perforated edges must be stopped and destroyed *tout de suite*; you pull to yank 'em out and you've still got all these small ripped pieces bound to the spine. Agh! And how about those multiply-paged magazine covers that are folded one under the other under the other like an unwelcome surprise accordion? HATE those! I'm ruthless, I have to be. Regular pages with ads on both sides? *Adieu!* I'm really good at this; I don't even need any special tools or anything. I can usually make a clean tear in any magazine's spine like nobody's business. The only publications that can pose challenges are *People*, *W*, and the *New York Times*'s seasonal *Style* magazines. *People* is stapled and once you get to the middle you can overtear, and then the whole magazine falls apart. *W* is oversized and so unwieldy and resistant that it's hard to get a good grip and clean rip. The *New York Times* is a pain because it's got the worst of the *People* and the *W* problems, as well as thin and unglossy (more defiant) pages. But, I prevail. My compulsion's so bad and involuntary that I even do it to other people's magazines. Like, at friends' houses or in doctors' offices or at the grocery store checkout line or at my hairdresser's—I'll just start tearing. It's terrible, I can't stop myself. People don't seem to mind, though. Mostly they just think it's kind of peculiar. Besides, when you think about it, I'm providing a service. I'm preparing your magazine and improving your reading experience by extracting the extraneous, and literally lightening it up. (Can anyone handle the full-blown, untreated September *Vogue*? Please. It's a thirty-pound doorstop. That thing could crush a baby.)

This new ad for COCO, however, I kept. It was good enough to tack up on my bulletin board. Its slogan: *Un nouveau parfum, une nouvelle femme.* A new perfume, a new woman. *Mais oui!* After so many years of N°19 I was ready for brave-new-world-old-world newness. I stared at the face of COCO, the quintessentially French Inès de la Fressange's face, in all its milky-skinned, dark-eyed, red-lipped, refined yet eccentric luxury. *Mon Dieu!* Inès practically glowed off the glossy page. Her raven hair was pulled back, revealing a pair of brushed gold button earrings embossed, *bien sûr*, with the classic Chanel logo, a pair of overlapping backward-forward *C*'s (for Coco Chanel). Her crisscrossed wrists, laden with gold bracelets and bejeweled cuffs, framed her chin, alongside quilted short black leather gloves. Her expression was mysterious, smoldering, contemplative. Was this the face that launch'd a thousand COCOs, and burnt the topless towers of Paree? Sweet Inès, make me immortal with a COCO kiss!

I had a crush on Inès for years. Even straight girls like me can get crushes on other girls every once in a while. It's a certain sensibility. Like tearing ads out of magazines, I can't help myself. Some women hold a fascination for me. They're beautiful or beguiling or smart or funny or quirky or talented or *some*thing; they possess that indefinable *je ne sais quoi.* That's how I saw Élodie. Élodie! I had to go see her! The ad said COCO was available exclusively at Saks. That *nouveau parfum* would be *parfait* for my upcoming blind date with the coke addict. Well, the criminal defense attorney who used to be a coke addict. Get it? Coke-COCO. God, sometimes life just *works.*

✦ ✦ ✦

Sonny, as I called him, was the son of a woman colleague whose family came from the deep South. She and I had become friendly in a maternal-filial way, and over lunch one afternoon she brought him up. Mothers of unattached grown children can be very cagey. Jewish mothers especially.

"He's a sweetheart, really," she said, taking a bite of her Cobb salad. "He's had some problems. Cocaine, mostly. But that's all in the p—"

"Mostly?" I said.

"Primarily, yes. There was a very, very brief heroin period but that was a—"

"Well he sounds just great," I said, mentally deleting him from my romantic Rolodex. "I hope he stays clean." I dug into my Chinese chicken salad, hoping I hadn't sounded overtly unkind to a junkie's mother. After all, it wasn't her fault her kid was fucked up. Was it? I'm like the least druggie person. How would I know what it's like? Coke was all the rage back then, but the good stuff you had to be rich to afford. It was Scarface, it was Eric Clapton and John Belushi—people I'd never know or be or date.

My colleague withdrew a wallet from her purse, extracted a photograph, and handed it to me. It was of her son the cocaine addict and his little boy. They were both lanky and strapping in their swimming trunks, squinting and smiling on a sunny beach. Dad was hot.

"I didn't realize you were a grandmother," I said. "He's adorable."

"My only grandchild or my only child?"

"Both. They look exactly alike. Actually, they look like

younger and older versions of Santino in *The Godfather. Godfather I,* of course."

"Who?"

"James Caan. Santino Corleone? Sonny? The curly hair? The physique?"

"Is that good?"

"You've never seen *The Godfather?* Dear. I'm speechless."

"A long time ago, I think," she said.

"Well, your son 'Sonny' is very cute. And his wife must be cute too."

"Well, they're divorced, so . . ."

"Living with anyone?" I asked, mentally un-deleting him from my romantic Rolodex.

"No. Interested?"

"Could be," I said.

That night Sonny called me at home. His voice was animated, virtually wired. He talked almost as much and as fast as I do, which is saying a lot. We laughed at how his mother had been our go-between. I heard him chewing. It sounded loud and thick.

"Double Stuf Oreos," Sonny said. "I replaced the junk with junk food. Gotta have the sugar at night."

"You're preaching to the choir, dear," I said. "I've never been into drugs, unless you consider Peanut M&M's and Snickers controlled substances."

"I think they'd be classified as Schedule II stimulants," he said, laughing. He began sipping something. "Milk. Good Oreos chaser. I'm about to open a bottle of water and smoke a Marlboro. You smoke?"

"Yeah. I'm as orally fixated as the next Jew."

"I really enjoy it. When I did blow I'd chain-smoke an entire pack at a time," he said, exhaling. "Everything's euphoric in the moment."

"You put your body through a lot," I said.

"It's a miracle I'm alive," he said. "All the doctors at all three rehabs said so. My former law partners, too. Right before they fired me. Excuse me, right before they let me go to pursue other professional opportunities."

"Oh my God."

"Yeah. Well, it was either that or face disbarment. I figure I got the longer end of the stick in the deal."

"You got arrested for drugs?"

"Arrested and convicted. Possession. All because I ran a fucking red light! Can you believe that shit? It was a routine traffic stop. Pulled me over, searched the car, frisked me. They found some coke. I'd just scored it, too. Pisses me off."

"You're a criminal defense attorney and you got busted for drugs, and then you had to hire a criminal defense attorney to represent you as a defendant who's also a criminal defense attorney?"

"I know. Crazy, right?" I heard him fire up a fresh Marlboro.

"Are you high right now?" I said.

"It's the sugar, ma'am. It's the sugar."

"Well, I guess you just like that white powder. 'There's a lot of money in that white powder.'"

"What?"

"What Sonny told his father Don Vito when he was trying to persuade him to get into the narcotics business. In *The Godfather*?"

"*My* son's the bomb. Lives with his mother. Can you believe it, we got pregnant on our honeymoon. In Paris."

"When in Paris . . ."

"It's cool. We didn't make it but we're good parents to our boy. After my trial—it's a disciplinary panel, actually—I took some time off to get my head together."

"You mean rehab?" I asked.

"Yep."

"So you have three heads? You said you went to rehab three times."

"Third time's the charm," he said.

"And what's a disciplinary panel?"

"It's a special committee within the state's Department of Justice that examines every complaint against attorneys. They review your status as a licensed attorney in the state—the state prosecutor's office filed the complaint."

A real live felon. A real live felon who'd gotten his law degree, I'd learned previously from his mother, from a real live Ivy League university. Its Latin motto had to do with truth and God and light.

"And then a lawyer for the state and my lawyer argued before the disciplinary panel," Sonny continued. "They filed briefs and stuff. The panel looked it over and issued a written opinion, like, a month later. The office of attorney ethics had set up a review of the case. It's like an in-house minitrial. Only they can disbar a lawyer. I was really lucky. Didn't have to do jail time. Lost my job—but not my license—and was ordered to do, like, three years of community service, pay a huge fine, and go to rehab. Best thing that ever happened to me—next to having my kid. I opened up a little private practice near my condo this summer. I can walk to work. Hired a sexy Brazilian assistant. It's going all right. But hey, let's talk about you. Mother says you're very pretty and bright. Are you?"

"I don't know."

My brain was swirling from all the verbiage. I had to process it. Sonny didn't seem to process anything; it was just boom-boom-boom, this happened, that happened, time for another cookie and a smoke. He was acquainted with The Dark Side. *My* Darkness was more along the "My creditors are breaking my balls/I'm getting fat/I feel lonely/I need slam-me-down-on-the-bed sex right now/ How will I ever become a published author if I'm a *Washington Post* Style section copy aide forever?" lines. The lingering ingénue in me imagined Sonny not trapped in court all day, but out somewhere shadowy and forbidden, scoring. He seemed like an outlaw, a dangerous, romantic, beautiful, doomed, and Really Tragic one. A cross between Al Pacino in *The Panic in Needle Park* and Katharine Hepburn in *Long Day's Journey into Night*.

Would Sonny ever substitute me for Double Stufs? Maybe. First I needed to score my own drug of choice: COCO. I was betting that it would give me a superpower that would give Sonny a sweeter high than Oreos or coke, and that the perfume from my dress would make him so digress.

"*It doesn't* matter if it's a criminal attorney, and depending on the state, it would have to be a crime of moral turpitude," Ray told me. "I would think a heroin and cocaine felony conviction would qualify as a crime of moral turpitude, risking disbarment."

"Yeah, but it was only cocaine and he wasn't disbarred," I said. "He was just canned from his law firm."

Ray was a friend of a friend who worked in my newspaper's Metro section. He was also a lawyer. Sonny intrigued me, no doubt about it. However, the phrase "trust but verify" kept running through my mind. This is what happens when you're a newspaper person, even if you're a mere copy aide. You're trained to think this way. Since there was no Google then—God, how did we ever survive before Google?—I'd resorted to an Actual Live Human for background. (Obviously I wasn't gonna ask Sonny's biased and likely denial-ridden yenta of a mother.)

"Well, there could be mitigating factors," Ray said, "such as first-time use, prior rehab or rehab begun, no priors, no other charges."

"Yeah, that's what it was. No priors." *Priors.* I love how these guys talk. It was like watching *Hill Street Blues.*

"If there are other charges such as trafficking," Ray added, "forget it. 'Cause unlike alcohol, coke and heroin are classified as controlled substances. Marijuana is somewhere in the middle. You should call the state bar where he lives for a more definitive response. But I think they'll confirm that it's not that clear-cut and that mitigating factors are weighed in. Does that help?"

"Oh yeah," I said. "He's not using anymore. He's all new now. He's a new man."

Un nouvel homme. And now, thanks to *un nouveau parfum,* I would become *une nouvelle femme* with that *homme.* A week or two later, the new man asked the new woman out. He'd fly up to D.C. for the weekend and stay in a hotel. Polite of him not to assume he'd shack up *chez moi.* A Jewish Ivy League junkie lawyer with good manners. We like that.

"You like French food and you like Paris," I told him. "I know this wonderful little bistro in Adams Morgan. It ain't exactly the City of Light but we can pretend."

My ridiculously darling dress, the one whose COCO perfume would be making Sonny digress, was a sartorial piña colada. I'd found it in, of all places, Loehmann's, and instantly fell in love with it. It had a Christian Dior "New Look" late 1940s/early 1950s silhouette with whimsical, ultrafeminine touches, as if the French *haut couturier* had gone off on a quickie little Maui getaway. The dress was a heavy cotton linen, very fitted in the bodice with a U-shaped scoop neck, adorable fruit novelty buttons down the front to the tiny waist, cap sleeves, and the kind of full, gored, below-the-knee skirt that swished when I moved and flared when I spun. The fabric was tropical citrine green, splashed with a colorful fruit motif including cherries, pineapple cubes, and coconuts that matched the buttons.

All I needed now was a tiny paper parasol to twirl.

Drink me.

Because it was such a busy dress and in a period style, accessories would have to be just right and kept to a minimum. Playful and fun but clean and simple. Something costumey and cute. Lucite cherry earrings and a matching bracelet? *Parfait.* Now, shoes? Oh, my beautiful cherry red suede platform stiletto slingbacks with open toes. I'd have to get crackin' on the pedicure. Purse? An adorable, structured red straw clutch with a red suede and silk blossom. And tons of Chanel's Gladiator, my best red lipstick. I ask you, how could any drug be better than that ensemble? Please, that ensemble *was* a drug. Ideal for the Indian summer we were having that fall.

I carefully folded the entire 'tude—dress, jewelry, shoes, bag, and all—inside a shopping bag to show Élodie when I went to get my COCO. On the drive over I thought of all the COCO-like words: Cocoon. Coca-Cola. Cock-a-doodle-do. Cockapoo. Cockatoo. Cocoa butter. *Cocotte.* Cocteau. Cuckoo. Cuckoo for Cocoa Puffs. Such happy words.

Cocaine? Not so much.

But, whatever.

"Élodie?" said a different saleslady, an American I recognized behind the perfume counter. "Élodie is gone, my dear. May I help you?"

"Gone?" I said. "Gone where, on Gitane break? Gauloise break?"

"On France break," she said. "The permanent kind."

"WHAT?!?"

"The boyfriend, he didn't work out. It was a couple of weeks ago."

"She didn't say anything? I'm her friend, Gigi."

"Oh, Gigi. Of course. Hold on. She left something for you."

She rummaged around and handed me a small Saks shopping bag with my name neatly written across it in that familiar Frenchy-French hand.

"Thanks," I said, confused and not a little wounded that my friend hadn't alerted me that she was—*bonjour*—LEAVING THE COUNTRY FOR GOOD. I never dreamed she'd take my advice! *Zut, alors!* I hate change.

"Something for you today?" asked the saleslady.

"Well . . . I don't know," I said reluctantly. I glanced at the attractive COCO display. Inès de la Fressange's beautiful face beckoned me. I'd been going to Élodie and only Élodie for more than ten years. It felt wrong and bizarre having someone else sell me my perfume. These are intimate matters. I took the COCO tester bottle and sprayed it on. Mmm. It was *fantastique*. Even more *fantastique* than the *Vogue* scent strip. I looked at the saleslady. She and I had no history. No intercontinental chemistry. No *je ne sais quoi*. I'd never have a crush on *her*.

Still, I wanted my COCO.

"It's nice, isn't it?" she said.

"It is," I said, nodding and reaching for my EVC (Emergency Visa Card). "It's really lush and rich."

"Shall I put you on the waiting list?" she asked. "We're sold out. All our stores across the country are."

"Of every version and every size?!? Of every *parfum* and *eau de*—"

"Every last one," she said. "It's been very popular."

"Samples?!?"

She shook her head and said, "Sorry."

I sprayed the tester into the air several times and walked back and forth through the COCO drizzle. At least I'd be permeated and my clothes and car would be. God, what a sucky day. No Élodie, no COCO, no nothin'. And Sonny would be here in three days and he'd already bought his plane tickets and my hula-hula cha-cha ooh-la-la outfit and I were all frothed up and ready to tipple into COCO!

Merde.

When I got home, I poured a stiff TaB on the rocks, took out my Parliaments, and opened the little Saks package. There was a folded note atop a mound of tissue paper:

Chère Gigi,

Please forgive me. Things were so rushed. I had to go. It's a long story. You were right, the old words young you said to me so long ago. You will come to Paris to see me sometime, anytime. Here is the address of my family . . . I will have a telephone number soon. This is my home and yours. The new Chanel parfum is beautiful for you, COCO. Use the Nº19 always, and the COCO for the rest of the times. (It is a delicious floral oriental for cold weather and night, but there are no rules!) COCO is French angelica, Bulgarian rose, Spice Island clove bud, Indian jasmine, Caribbean cascarida, frangipani, mandarin, orange blossom, Tonka bean, sandalwood, leather, wood, vanilla, and mimosa.

They say the scent is meant to evoke the gilded mirrors, Venetian chandeliers, and lacquered Chinese screens Mlle. Chanel had in her apartment. You will see all these pretty things and many more when you come to Paris. I hope it will be very soon, mon petit chou.

Affectueusement,
Ton amie toujours,
Élodie

P.S.: The store sold all the COCOs! Voilà, what I gathered and hid for you. xooxx

I unwrapped the tissue paper mound.

Chère Élodie:

Sacré bleu! Seventeen glass vial samples of COCO!?! SEV-ENTEEN?!? Élodie! You gorgeous Gallic goddess! I bow down to your fragrant Frenchness forever. Merci bien un mil-lier de fois ou au moins dix-sept mille. [Thank you a thousand times or at least seventeen thousand.] Naughty of you not tell-ing me you were leaving or why (it's a French thing, I know; I wouldn't understand), but knowing toi I'm sure you'll land on your Chanel spectator slingbacks and never once smudge your red lipstick.

I'll miss you, Élodie. Your impeccably kohl-rimmed eyes. The musical sound of your voice. Everything.

Amour toujours de ton petit chou,
La Gigi xoxoo

✦ ✦ ✦

Snowflakes swirled, confetti sparkled, and tiny cherries bobbed all around the Eiffel Tower, the Arc de Triomphe, and Notre-Dame as Yves Montand wistfully sang *"Le temps des cerises,"* a bittersweet song—a poem, really—from the late 1800s. It's about love in the time of the cherries. Montand cautioned his fellow men, *Quand vous en serez au temps des cerises, si vous avez peur des chagrins d'amour, evitez les belles.* When you are in the time of the cherries, if you are afraid of love's sorrows, avoid the beautiful (women).

"It's extraordinary," I told Sonny. "Cherries *and* snow?!? 'Cherries in the Snow,' aw. The old Revlon lipstick. Where did you ever find this?"

"Just a little present for ya," he said. "Like it?"

"It's the most amazing snow globe-slash-music box I've ever seen," I said. "God, it even matches my outfit!" *Snow* globe. Snow. Blow. Was *that* why the sexy nose monster bought this for me? A punny little co*caine* keepsake? (Actually, it wasn't little at all. The thing was the size of a Waring blender.)

Saturday afternoon in Sonny's nondescript D.C. hotel room. I'd told him to dress comfortably and casually; La Fourchette, where I'd made us dinner reservations, serves excellent classical fare, yet it's a come-as-you-are establishment. Sonny was wearing what he wore on the plane: a short-sleeved Lacoste polo shirt with green, navy, and white stripes (plus the obligatory green Lacoste crocodile), a pair of faded Levi's, and navy nubuck Bruno Magli moccasins with rubber soles. He looked *good.* Terrific body. Sinewy. I'm telling you, Santino's thirtysomething doppelgänger.

I too was a vision. I was a piña colada. A COCO sample-scented one. (*Merci,* Élodie!) I was sitting with my legs and bare feet tucked under me on the king-size bed, the voluminous skirt of my dress rounded out all around me like an enormous unfurled garden umbrella. Sonny leaned in and kissed me. Not a big-deal COCO *l'amour* kiss; more of a friendly domestic howyadoin' sort. He lit a Marlboro and took a sip of his Diet Coke. He devoured a Double Stuf (I'd brought him a whole box), kicked off his moccasins, and sprawled out next to me, lifting the fabric of my dress so as not to wrinkle it. He perched an ashtray on his stomach and offered me a drag of his cigarette. I took a puff, watching his tanned feet twitch.

"What time are we leaving for dinner?" Sonny asked.

"We've got two hours," I said, rolling off the bed to smooth my dress. "Twooo whole hours to kill. What oh what shall we ever do?"

"Come 'ere," he said, putting the ashtray on the nightstand and pulling me by the hand back into bed next to him. "That's some dress you got there. If I put a straw in your lips . . . could I drink you?"

"I like your crocodile."

"Yeah? What about me? How'm I doin'?"

"Well, Mayor Koch, I'd say your Parisian snow globe-slash-music box has made *beaucoup* waves." Sonny probably didn't get the reference to the New York City mayor's public mantra, but it didn't matter: he kissed me passionately. Though he'd said it was the dress and hadn't mentioned my hypnotic aroma, I knew it had to be the (subliminal) COCO power. He kept kissing me as he unbuttoned my cherry. My pineapple cube. My coconut. I was one excited piña colada. Thank God I'd worn the good cherry red bra and matching panties. Sometimes a gal's gotta be coordinated all the way down under. (Admittedly, it might've been a *tad* hasty to be showing the boy my knickers—this was, after all, our first

in-person meeting—but I'd been talking to him on the phone for
weeks, so that's, like, *hours* of virtual dates.) Sonny got undressed.
His wiry body was taut and muscular. Over me he felt stiff and big
and flexible as a diving board.

And then, and then, and then . . .

Pffffft.

I waited for the silken alcoholic berry richness of the Pinot
Noir to kick in and take the edge off the lingering awkwardness.
Sonny and I were seated by the desirable front windows of the
crowded, cozy La Fourchette, studying our menus and not talking
to each other. Our cute Algerian waiter, Marcel, was short, dark,
and handsome. He seemed a lot more appealing at the moment
than my date. Not a good sign.

"*Je commencerai avec la salade vert avec la vinaigrette et le from-
age de chèvre chauffe*," I told Marcel, just as I'd told him or one of
his attractive coworkers the previous 359 times I'd been to the
restaurant. "*Et alors j'aurai le filet mignon au poivre, puit-cuisiné, et
les batons frites de pommes de terre.*"

"*Très bien, mademoiselle*," he replied. His thick, jet black hair
gleamed blue in the honeyed candlelight. "*Et pour vous, monsieur?*"

"I don't speak French," Sonny said.

"*Il ne parle pas le français*," I said. He doesn't speak French.
"*C'est bien dommage.*" It's quite a shame.

"What may I bring you, sir?" he asked Sonny, smiling and
winking at me.

"Well, since we're pretentiously pretending we're in Paris, I'll
start with the vichyssoise—it's served cold, right? Ice cold? Potato and leek soup?"

"*Oui, monsieur,*" said Marcel.

"And then I'll have the roast lamb with white beans. *Haricots blancs.*"

"Very good!" I told Sonny. "*Haricots blancs.* You go, white bean man."

"This is a nice place," Sonny said, looking around. The restaurant's walls are covered in murals. Café scenes.

"Yeah," I said, lighting a Parliament.

"I've been to D.C. before, to visit my mother. She never brought me here."

"Well, it's kind of a romantic place."

"Speaking of which . . . Uh, what happened? Back at the hotel?"

"It's okay," I said insincerely.

"Okay, look, I did some blow before I got on the plane," he said, lighting a Marlboro. "I know what you're thinking. It's not like that. I'm just chipping. I got nervous about the trip."

"You're still using?" I said. I couldn't believe it.

"*Chipping.* Not using. Not really. Anyway, you *think* you're hard, you get horny as hell. But you're . . . you're really not." He stroked my arm. "Don't be upset." He rolled a cherry on my bracelet between his fingers. I drew my arm away to pick up my wineglass.

"Why are you here?" I said.

"Why not? And I wanted to please Mother. She's constantly trying to set me up with people. Girls she thinks will 'save' me with their . . . their *goodness.*"

"If she asks, I'll say the chemistry was off, okay?" I said.

"Sure," he said. "It's not as if she hasn't heard that one before."

Marcel arrived with our appetizers. He lightly brushed up against me while he ground black pepper on my salad, and left.

"That time we were talking and I asked if you were high?" I said. "Was it really from the sugar in the Oreos?"

"I don't remember."

"Do you even *like* Oreos?"

"Mmm, I like this soup. It's great. Want some?"

My shrink always told me that my forte is romanticizing *caca*. Sonny didn't consider himself a real junkie because real junkies don't wear Italian designer moccasins and crocodiles on their shirts. Do they. Well. *Another dream over the dam*, as Joni Mitchell sang. This time I'd only wasted about a month. That's not too bad. I looked out the window. Saturday night lovers strolled past, arm in arm. They were laughing and talking. They looked so high. Was it love or blow? I lay my nose on my wrist and inhaled. It was still COCO'd. At least I had that. And I had my sweet memories of Élodie and fourteen glass vial samples of COCO. I considered leaving Sonny there, with his vichyssoise and delusions and faux hardons. But I'm not that kind of girl. I'm a pragmatist at heart: Marcel and his beautiful coiffeur hadn't yet brought me my *filet mignon au poivre*, and La Fourchette's *mousse au chocolat* really rocks.

The end? Almost. I dropped Sonny off at his hotel. He kissed me on the cheek and said, "Good luck with everything. This was fun." That was one word for it. I could think of another word. Actually, I could think of seven thousand. Mostly what I was thinking was, Thank God I took the snow globe-slash-music box with me before we left for the restaurant because I can't wait to get home and catch *Saturday Night Live.*

As I approached the Carter Barron Amphitheatre at the intersection of 16th Street and Colorado Avenue in Northwest, the yellow light was juuust turning red. I gunned it. Suddenly there were flashing lights in my rearview mirror and I heard the siren.

Merde.

I pulled over and rolled down my window. A very attractive cop who looked like a ripped Ken doll came over. He asked me for my license and registration, and if I was aware I'd run a red light.

"I know, officer," I said. "But I just had the worst date of my life! The guy's a fucking cocaine addict, can you believe that shit? And I bought this COCO and—"

"Please stay in the car, ma'am."

I nervously lit a Parliament and popped a TaB. (I always keep a can in the car. Room-temperature TaB is not a problem for me.) From my rearview I saw there were actually two cops. Let's call them One and Two. (One is the Ken doll.) They were talking to each other. They walked back to my car. Two shined a flashlight on me and the seats. One said, "Miss Anders? Are you aware that you have $379 in unpaid parking tickets?"

"I do?" I said in my most *shocked, shocked* tone. "How awful, dear."

"Ma'am, we're gonna need you to step out of the car, please."

"You are?" I said. "Why?"

"Please step out of the car," he said. Then to Two: "Go ahead. Call for backup."

"WHAT IS GOING ON?" I screamed.

Dozens of cars were clogging the two-lane street, which had now become a single lane. People were slowing down to gawk at me like I was a criminal. A piña colada criminal with adorable Lucite cherry jewelry, but a criminal nonetheless. Great. Great-greatgreat. Another police car pulled up behind the first one. Two women officers got out. Let's call them Three and Four.

"Sorry, ma'am, but I'm gonna need to frisk you," Three said.

I couldn't help but laugh. "You need to WHAT?"

"Turn around and spread your arms and legs, please."

At least they all said "please" a lot.

"We're sorry about this, but an infraction's an infraction," said Four. "You need to pay your parking tickets and stop running red lights. We just had a college professor with the same—was he from Georgetown University?"

"Yep," said Three, patting my hips.

"And he was none too pleased, either," said Four. "Started screaming 'bout 'Why are you guys hassling ME? This is what I'm paying my freakin' taxes for? Why aren't you out chasing druggies?'"

One, Two, Three, and Four howled at that one. Yeah, it was a scream, all right. The word *Kafka-esque* came to mind. Hold it. *Chasing druggies.* Isn't this EXACTLY what happened to Sonny? Running a red light? *Merde! Merdemerdemerde!*

"She's clean," Three told One. "She smells good, too. You got the cuffs?"

"CUFFS?!?" I screamed.

"Sorry, Miss Anders," One said, locking the bracelets with two clicks. "It's the law. We have to take you in."

"YOU'RE ARRESTING ME?!?"

"Yes, ma'am. And we're impounding your vehicle. You can pick it up after you pay your fines." He sniffed. He sniffed again. "Hey that's nice, what you're wearing. What perfume is that?"

"COCO," I said, praying to God this wasn't some sick sex con.

"You're wearing hot chocolate?" One said.

"NO," I said. "COCO. COCO is the name of the perfume. CHANEL. *HELLO.*"

"Hey guys, come over here," he said. "You've got to smell this. You mind?"

"I'M IN HANDCUFFS," I screamed. "WHAT DO YOU THINK?"

Two, Three, and Four clustered around and smelled me. The consensus was COCO *loco* love, especially from One. It's always nice when a wonderful evening ends with a compliment from the officer who's arresting you.

"Thanks," I told him. "Are you single, by any chance?"

Le Tank

I had a preppy. I know. *Me.* With a *preppy.* On *purpose.* Every gal should acquaint herself once (and believe me, once is more than enough) with a member of that theoretically elite subspecies. It's like "slumming" in a world of Dewar's, Docksiders, and duck decoys (preferably mallard). It's very interesting. It's like being a tourist in L.L.Bean and Brooks Brothers Land. I attended a private prep school before college, or should I say college*s*—I went to three different ones, including a stint abroad—so it wasn't as if I was unfamiliar with that whole Holden Caulfield genre. I normally repel the type. You know, Ivy- and frat-bound Gentile boys, gangly jocks in all-natural fibers, with pronounced weaknesses for too many "brewskis" and straight-haired, straight-bodied girls whose personal grooming and relationship to makeup involves Ivory soap and ChapStick. Exclusively.

Superficially, my preppy, whom I teasingly called Dr. Ivy, was no different. He was just older. That's when they start with the Eau Sauvage aftershave, have turned exuding the eternally boyish, cocky nonchalance into an art form, and have mastered the uncanny ability to color-coordinate an endless collection of striped grosgrain ribbon watchbands with an equally endless collection of matching brass-looped striped grosgrain ribbon belts. But. Underneath the usual trappings, Dr. Ivy was unusual: he was attracted to me. To my Vargas girl red lips and bubblicious extroversion and curly hair a fact no preppy before or since has ever been. And can just *any* prepster recite entire Emily Dickinson poems and F. Scott Fitzgerald passages by heart? Or be a cross between Alistair Cooke and Johnny Carson? Or, when you're in bed together, call yours a "transcendental sensuality"?

Also, I loved his Midget. It was irresistibly cute. It made me feel like Daisy Buchanan whenever I rode in it with him because it looked like one Jay Gatsby would've owned: a dark British racing green 1963 MG Midget convertible with retractable frog-eye headlamps and twin burnished leather seats the deep caramelly color of weathered Louis Vuitton leather trim. Dr. Ivy's perfectly restored pride and joy, it was in what he called "mint condition." I guess that's true—as long as you didn't expect it to actually, like, move that much. It was an auto but it wasn't reliably mobile. Curiously, it roared along fine whenever I was in it—the Midget loved me. But when I wasn't, it was constantly breaking down and stranding Dr. Ivy in inconvenient places at inconvenient times. He'd call me, frequently drunk, at my dorm at all hours to come get him. At first I did it. I liked him, it seemed like a cool thing to do because it made me feel important. After the fourth or fifth time, though, the routine got old and I'd yell at him to call goddamn AAA. Then he'd yell at me that he wasn't a goddamn member. Then I'd yell at him to contemplate the goddamn phrase "learned helplessness." Then he'd yell at me that I was being goddamn unfair. Then I'd yell at him to call up some less goddamn unfair student groupie to come rescue her goddamn English professor and his goddamn English Midget 'cause I was goddamn sick of it. Then we'd both start laughing and make plans for the weekend. Using my Honda Civic.

How much time romantic runners-up take from your life, time that can never be recovered, was the subject of a short story I read in college in *The New Yorker*. I don't recall all the details, but I do recall the author was a woman and her piece sounded suspiciously roman à clef-ish. It was about this Harvard undergraduate who was having an affair with one of her English professors. They'd just had an assignation and she was walking back to her "house," as the Harvard dorms are called, in the Cambridge rain. Her high heels clicked on the campus's wet gray stone walkway. Click-click-click. And as her shoes clicketty-click-clicked, she pondered how much longer this affair might last, how many more unrecoverable moments of her life it would consume and displace.

Time, time, time, Simon & Garfunkel sang, *see what's become of me while I looked around for my possibilities. . . .* Some lovers take up weeks, months, or years. If you're in college, you mark time by semesters. And you don't need to be in the Ivy League or published in *The New Yorker* to have an affair with your professor. You can, as I did, do it anywhere. Think *The Way We Were* and a touch of *Love Story*, except I was an undergraduate English major and Dr. Ivy was one of my professors. He'd graduated from an Ivy League university, a fact he wasted no opportunity to name-drop. He always wore a gold signet ring bearing his alma mater's crest on his right ring finger, presumably to remind himself and the rest of the world that he was "worthy," and that our institution, which he regularly referred to as "a monument to mediocrity," was *un*worthy of *him*.

"If this place is such 'a monument to mediocrity,' " I once asked during class, before we'd become officially *intime*, "then why are *you* here?"

"*Noblesse oblige*," he said, smiling.

"Harvard wouldn't have ya, huh?" I said. That made everybody laugh. "Well, *tant pis* [too bad]. Harvard wouldn't have me, either. Guess we're even. Evenly mediocre."

"We may not be at Harvard," he shot back, looking at me intently from behind his small round tortoiseshell eyeglasses, "but I expect a lot of you. And I'm going to get it."

"*What is* going *on* with you two?" asked my classmate, Dana. She was a virginal, vicarious brunette majoring in American Studies. She was taking Dr. Ivy's class as an elective because she planned to go to law school and thought that learning how to read and write might actually be, like, useful in that field. We always had long conversations after class, mostly about our popular professor. "It's so funny. You toy with him. He's twice our age and you—"

"I toy?" I said. "Really?"

"You toy. You toy and torment him."

"I do not. Wake *up*, dear. We're just flirting."

We were sitting outside on a bench on the campus mall as Frisbee players tossed and jumped and slid across the green on a gorgeous October afternoon. My favorite time of the year. October was also the name of one of my and Dr. Ivy's favorite Robert Frost poems. *O hushed October morning mild,/Thy leaves have ripened to the fall; . . . Begin the hours of this day slow./Make the day seem to us less brief./Hearts not averse to being beguiled,/Beguile us in the way you know.*

"Well," Dana said, "you're not intimidated by the guy."

"Why should I be intimidated?" I said, exhaling my Parliament smoke. "Just because he went to—"

"Oh, you're just too glamorous with your Hollywood red lipstick and your cigarettes and TaBs."

"It's not Hollywood—that's cheap display. It's Chanel—that's *amour*, mama."

"You're not thrown, not even a little bit?"

"What's there to throw? He's not much bigger than his car. His car's called a Midget—hello."

"Too bad he's our teacher, right?" she said. "Because smart women like intelligent men. Intelligence is crucial."

"So's a big ding-dong. This is why Dr. Ivy's no big deal. He's what, two feet tall? Please. The guy's a Munchkin. He's the Pillsbury Dough Boy. My French boyfriend, Bub, he was at least five inches taller than Ivy, and the French aren't exactly jolly green *géants*. Except for that scary Gérard Depardieu guy—what's up with him? Anyway, I think Dr. Ivy's got a definite Napoleonic complex."

"He's not Napoleon," she said. "Dr. Ivy's comfortable in his skin."

"Okay, so he's not *poison* Ivy," I said, reaching for my ever-present can of TaB. They used to sell it in soda machines back in the civilized Pleistocene Era. "And he's not tall, either. Not in that special, special way. Not that I would know firsthand. Or second."

We giggled. A falling red deltoid leaf strayed into my hair. Dana picked it out and handed it to me. "Here he is," she said of the leaf. "Dr. Ivy. He's dropped in to see you."

"*Merci*," I said, turning the stalk in my fingers. "That's *Docteur Lierre* to you. Doctor Ivy in French. And see, *he* comes to *me*. I don't have to run after him."

"Ah yes, your preppy French lover."

"He may *have* gall but he's no Gaul," I said. I knew from whence I spoke.

"He almost has that patina of old money, that tattered elegance," Dana said. "But in reality he probably doesn't have it."

"Dear, didn't he say he lives in . . . where is it? Way out in the sticks? Like thirty MILES from here? In a whole other STATE? That means he's poor. A poor poseur. He poses poorly, that's what it is. He's bluffing, I can tell. There's something good ol' boy about him. Confederate. He's trying to convince everybody that he's this intellectual New England Protestant with his 'I'd Rather Be Sailing on the Cape' key chain. But that's Fear, not Cod."

"There's preppies below the Mason-Dixon Line."

"I'm talking rednecks. I'm talking deeep, deepdeepdeep Dixie, you know? Red dirt road Evangelicals."

"Evangelicals," Dana repeated, skeptically.

"Okay. Not Evangelicals. But Baptists. And he thinks he's above it because he spent four years drinking beer and perfecting his French *bon mots* and getting laid in this idyllic place with an endowment in the billions. As opposed to, you know, here. Here it's, like, Frisbee U."

"Well, *you* think *you're* above it because *you* spent a semester drinking wine and perfecting your French and getting laid in Paris by—"

"Yeah. But I'm the real thing. *La vrai chose.* I'm not faking anything."

"Certainly not orgasms," Dana said, laughing.

"*Jamais,* baby." Never.

✦ ✦ ✦

Dr. Ivy may have *appeared* to be to the manor born (a custom more honored in the breach than the observance, as Hamlet said), but I had what I considered worldlier credentials. I'd lived abroad in Paris, France, thank you very much, and that's way better than any Ivy League education to *moi*. I'd whiled away a whole spring and a summer and part of an autumn with all manner of Parisian-born denizens. My favorite was an adorable *sommelier* named Robert (pronounced Raw-BAHR) who wanted me to call him Bub. I'd read Baudelaire's *Les fleurs du mal* (Flowers of Evil) in the Jardin des Tuileries. Most significantly, I'd fallen in love. On 13 Rue de la Paix. Not with a man. With a tank. *Le* Tank. Le Tank de Cartier. Le Tank of all *temps*, tank you very much. *Quel* watch! The rectangular—or was it square, or was it both?—white face framed in 18-karat yellow gold. The Roman numerals in black. The pale silver opaline dial. The sword-shaped blue steel hands whose color matches the octagonal sapphire cabochon in the winding gold crown. The substantial black alligator strap with the 18-karat yellow gold buckle. Stunning. Elegant. Simple. Intelligent. Chic.

It cost more than—and I figured this out—two packs of Swisspers cotton puffs, 450 six-packs of TaB, and one breast aug'. I was a poor teenybopper wearing a Seiko with a round white face and a stainless steel-gold tone chain-link bracelet. It was a high school graduation gift. I wore it with everything. Now I was dreaming of expatriation and being seduced by a killer French Tank in Cartier's window near the Place Vendôme and The Ritz. Where Hemingway holed up. Where F. Scott Fitzgerald holed up. Where Coco Chanel holed up. Did I really belong here? I took off my sunglasses to have a better look at Le Tank. I'm pretty nearsighted as it is, so I pressed my face up against the glass, whereupon a well-dressed saleswoman named Gabrielle de la Quelque Chose (I made up that last part; it literally means "Something or Other") tore out of the store bearing a spray bottle, a cotton rag, and a withering French scold.

"Je suis désolée, madame," I said in my best prep school French. *"Mais je ne peux pas l'aider. Je suis dans l'amour! Cette montre-bracelet est si belle! Là-bas? Avec la bande noire d'alligator? Aussi, je suis myope."* I'm sorry, madame. But I can't help it. I'm in love! That watch is so beautiful! Over there? With the black alligator band? Also, I'm myopic.

That made her smile. Gabrielle invited me inside to check it out. I wasn't sure I should; it can hurt to be so physically close yet so financially far from an object of such dreamy desire. On the other hand, I was in Paris, and Paris is nothing if not dreamy desire. Which Tank in which collection of seven different styles of Tanks would I care to see? And in which of the five sizes, from minuscule to ultra-large? My abnormally large American head spun from the luxurious *abondance française*:

- Le Tankissime? (Yellow, white, or pink gold chain link, with or *sans* round-cut white diamonds on the long sides of the face. Strictly optional timepiece; if you own a Rolex or even a humble Seiko like mine I doubt you'd ever covet this to the point of sleep deprivation.)

- Le Tank Divan? (Large rectangular face turned sideways, so that its shape is exaggeratedly wide and squat. Some are framed all around with tons of little white diamonds. Extremely broad band. Limited edition. Over-the-top blingy. What a rapper would wear to look understated.)

🎖 Le Tank Française? (Caramel-colored alligator strap with a yellow gold frame. Or classic black alligator strap with a white gold frame. Or, in a total departure from the alligator strap motif, a steel bracelet with a pink mother-of-pearl face. *Beaucoup* variations on the theme. But none made the Earth move. And for these prices, sorry, the Earth really must move.)

🎖 Le Tank Américaine? (Skinny, long, slightly convex, vertically stretched out rectangular faces. Feminine. But no cigar.)

🎖 Le Tank Chinoise? (Pink gold, diamonds, black silk strap. Nice for all those black-tie occasions with Chinese emperors. I have so many of those. But, as post office clerks say when there's a line, "Next!")

🎖 Le Tank Louis Cartier? (Caramel strap with yellow gold or black strap with white gold. No and no. This was actually starting to get easier. You eliminate what you don't want in order to find what you do. Very much like men.)

🎖 Le Tank Solo? (COME TO *MOI, MON AMOUR!*)

When I spot exactly what I want I'm a flying black crow diving down to pick up something shiny. I hastily removed my Seiko and fastened Le Tank Solo around my tiny wrist. Ahhh. Yesss. Cartier, Tank me. I gazed at it from different angles. I gazed at it on me in a mirror. I waved to invisible commoners as I'd seen Jackie O and the newly engaged Lady Di do, wearing their Tanks. An amused Gabrielle proceeded to explain the famed watch's *histoire* as I sat on a leather-covered chair in the hushed, cozy, golden, chandeliered *maison*, continuing to wave. Before Louis

Cartier designed MY Tank in 1917—he was one of the jeweler grandsons of founder Louis-François Cartier—everyone wore pocket watches. Louis Cartier wanted to honor the American troops in the First World War. He based the design of MY Tank on the modern, horizontal lines of the French Renault military tank. As a tribute for his help in the war, Louis Cartier gave the first Tank prototype to General John Joseph "Black Jack" Pershing, commander-in-chief of the American Expeditionary Force in Europe. Besides its unkillable splendor, what's extraordinary about Le Tank was and is its sleek unisexuality. It has to do with proportion. The large size looks right on a man, the petite one looks perfect on a woman. I think I'd wear the medium size. It's fabulous any way. Soon everybody wanted one. (And, if everybody had the money, everybody could have one.)

"*Et voilà*," Gabrielle concluded. Which is to say, There it is.

"It's the most beautiful watch I've ever seen," I told her, sighing. This rarefied atmosphere, this civilized store, my cultural learning experience—was a whole new world. Why couldn't this academic term go on forever? I was just starting to get the hang of using bidets and wearing head-to-toe black and sprinkling brown sugar on plain white yogurt for lunch. I reluctantly unfastened MY Tank from my wrist and returned it to Gabrielle, feeling as if I'd just been forced to take off all my clothes. Wearing Le Tank was like wearing an entire designer outfit. "It's a dream," I said, barely holding back my tears. "It's a dream. Someday." *Someday when I'm awfully low/when the world is cold/I will feel a glow just thinking of you. . . .*

"*Retourniez quand vous voulez, mademoiselle.*" Come back whenever you like. Gabrielle said, kissing me on both cheeks.

So I did. I stopped in to see Gabrielle at Cartier every week for months to visit MY Tank. I thought of it as a kind of petting zoo. Maybe I couldn't own a real Tank but I could own a, quote, Tank, unquote. Mick and the Stones got it right: *You can't always get what you want but if you try sometimes you might find you get what you need.* On weekends, I began roaming the City of Light's flea markets, *les marchés aux puces,* searching for the perfect Tank, what the French call a *"faux authentique."* A believable knock off. Before long I became a virtual Ph.D. in Parisian flea markets, some of which have been around since the 1700s. They're all fabulous and fascinating, like their names: Aligre, Beauvau, Montreuil, Porte de Vanves, Saint-Ouen. I once saw several credible, quote, Louis Vuitton, unquote, suitcases, not that I'd ever want them. If you don't have a battery of personnel to schlep your luggage, why in the world would you want suitcases that weigh a ton before you've even put anything inside them? I've had the same black nylon LeSportsac "Extra Large Weekender" for years and it shows no signs of flagging. I've even washed it several times in the machine, and it still looks brand new. It weighs nothing and is strong and huge enough (16" x 24" x 12.5") to hold—and this is really saying something—*aaall* my toiletries and makeup and hair products; blow dryer and diffuser attachment; Parliaments in the hard flip-top box; jewelry; and backup underwear, tank tops, and TaBs.

After months of Saturdays and Sundays at *les marchés aux puces,* I never found a single Tank. I did, however, find, fall in love with, buy, and never take off a beautiful, tiny, *authentique* but not *faux* enamel pendant charm of the Basilica of Sacré Cœur in Montmartre. My rabbi, Bruce Kahn, says that God's power is all around us to utilize; it's everywhere. All we have to do to tap into it and connect is to invite God in. Since Jesus was Jewish and Sacré Cœur is a major house of worship located at the very highest point of Paris, I figured it would be okay to have them stand in. I mean, I was out of town and everything. Wearing my charm made me feel that *Dieu*, God, who is multilingual, was with me, watching over and silently guiding me toward that evasive Tank.

Rain. Sheets and sheets of it. The bleak weather underscored my mood. On my last day in Paris, it was storming and very cold, with biting air. I thought of metronomes' pendulums watching Bub's windshield wipers sweep away the torrents. Bub was my *sommelier* friend and sometime lover. He was driving me to Charles de Gaulle Airport while I failed to not sob. If I let my sobbing go on too long I get hiccups, those really embarrassing, painful ones. So I distracted myself by thinking of what Hemingway said: "If you are lucky enough to have lived in Paris as a young man, then wherever you go for the rest of your life it stays with you, for Paris is a moveable feast."

"*Prends-la, chèrie,*" Bub said, handing me his filterless Gauloise. Take it, darling.

I hated French cigarettes but I was so heartsick about leaving French France and French Bub *and* never having found that French Tank that I smoked it. It was *très* disgusting. But in a good way.

"Please don't come inside with me, Bub," I said as we pulled up to my terminal's curb. Like his city, Bub was a romantic; I knew it would only exacerbate this agonizing airport *au revoir* if he sat with me waiting for my plane. Even though it couldn't have been more than 40 degrees out, Bub broke a sweat handing a porter my 783 bags. He took me in his arms, held me tight, and kissed my mouth as I clenched the lapels of his blue raincoat, wrinkling them.

"Ne m'oublies pas," Bub whispered into my hair. Don't forget me.

I tried to smile. The harder I tried, the harder I cried. Bub touched my wet face. Raindrops clung to the tips of his long black curls, they streaked down his tawny cheek. Bub's was a long, lean face, a very French face, with crinkly brown eyes and high cheekbones and a full lower lip. This was the face I was willingly leaving.

Here come *les hiccups.*

Bub pulled a thin rectangular gift box from his raincoat pocket and put it in my hand. It was covered in a lovely forest green and ivory fleur-de-lis wrap and tied with a red satin ribbon.

"Un petit cadeau pour un bon voyage," he said. *"Ouvres-le après que je parte, chèrie. Avant que je commence pleurer et hoquer aussi."* A little gift for a good voyage. Open it after I leave, darling. Before I start crying and hiccupping too.

Bub went back to the car and returned with a small bottle of Perrier. I drank it, holding my breath. That's the only cure for hiccups. I kissed and hugged Bub for the last time, squeezing his long, soft hands in mine before letting go. I watched him drive away in the rain in his gray Renault. I couldn't remember ever feeling so *tragique.* I tucked the pretty box in my satchel and went inside. Somewhere over the Atlantic, I started to unwrap it, then changed my mind. I could wait. No I couldn't. I tore the thing open. *Sacré bleu* and *Sacré Cœur!* It was a Tank! Le Tank Solo! The ultimate *faux authentique!* Not too big, not too small. It was PERFECT! The Tank that had eluded me for months and months and months! After so many fleas and *puces* and far-out Métro stops across Paree! God lives! Even in France!

There was a little handwritten note in Bub's Gaul scrawl:

Gigi, Gabrielle de Cartier connaît un gars qui vend ces Tanks. Un américain de Los Angeles, peux-tu le croire? P.

[Gigi, Cartier's Gabrielle knows a guy who sells these Tanks. An American from Los Angeles, can you believe it?]

✦ ✦ ✦

Good thesis and intelligent commentary. Although (as you have no doubt seen from our class discussion) you are wide of the mark in some of your assumptions, your analysis is adroit, even perceptive at points. It's also the best—so far—I have gone over from the class. A–.

Dr. Ivy had spoken. Well, he'd commented, in his tiny, barely legible idiosyncratic script, on my first term paper.

"You have good taste in students," I told him after class. I casually ran my fingers through my hair with my left hand, hoping he'd notice what was on my wrist. My magical Tank. It was the first time I'd worn it in public. Surely it would roll over and capture this American preppy. My soon-to-be POW. Prisoner of Wantonness.

"I think I have good taste in students, too," Dr. Ivy replied, smiling. He took his old tweed jacket off the back of his wooden chair and tossed it over his shoulder. "You got time for coffee?"

"Sure."

"Nice watch you got there," he said. How do you say GOT-CHA! in French? Haaa. "What're you, rich?"

"Oh, frightfully," I said with a little snicker.

We walked to the sanctum sanctorum, the English Department faculty lounge, where he poured us stale black coffee into white Styrofoam cups. Not exactly *café crème* at Café Le Flore on Saint-Germain-des-Prés with Bub. God, American reentry was a bitch. A couple of my other professors wandered by, saying hello and looking vaguely puzzled. I waved at them like Jackie O and Lady Di.

"You remind me of the green light in *Gatsby*," Dr. Ivy said, stretching his short legs and turning wide open toward me on the long brown ratty sofa. As usual, he was sockless. Today he was wearing ancient brown penny loafers whose tops were secured to their bottoms with three-inch-wide strips of silver duct tape. Classy. Apparently, this is what all the preppies do when their loafers begin falling apart. They like old stuff. If they have new stuff they want it to look like old stuff ASAP. Apart from, I surmised, girlfriends.

The green light in Gatsby. How many impressionable little girls had that line worked on? Some students, like my pal Dana, look up to their teachers as if they're rock stars. I'd pretty much outgrown that shit the minute I saw Dr. Ivy's duct-taped shoes. I had a little crush on him, of course, but he was, at the most, domestic rebound material *après* France. Wasn't he? He was one slick preppy, I'll give him that. He had that Bill Clinton catch: charismatic and bright and fun, but with a stupid personal life. Big head thinkin' with the little head. Maybe that would explain Dr. Ivy's divorce, not to mention the duct tape.

In *The Great Gatsby*, Daisy Buchanan said she hoped her daughter, Pamela, would grow up to be a pretty little fool. "That's about the best a girl can hope for these days, to be a pretty little fool." Would I become a Pamela if I went with Dr. Ivy, who only seemed capable of dealing with the very last thing that happened to him? Could I take him seriously? Did I need this in my life? Maybe, maybe, and maybe. I wanted to have fun and retain my independence. This is one of those positions you can't simulate. Men may be obtuse about a lot of things, but a woman's independence isn't one of them. They can smell it. It's a pheromone.

"What about the green light?" I asked.

"Oh, the go-light Gatsby picked out at the end of Daisy's dock," Dr. Ivy said.

"Daisy go-lightly? Holly's sister?"

He laughed and said, "You're really refreshing."

"Thank you. I am a TaB."

"Seriously, what's your story?" he said, lowering his voice and quickly looking around the room. "Because I've been sensing for a month or more that there's something between us. Or that there could be. I mean, I'm unattached and you're—you're not seeing anybody, are you?"

"You're picking me up in the faculty lounge?" I whispered.

"Not literally. But I would like to pick you up and throw you in my bed."

"I have to go," I said, looking at my Tank. "I have a class. Rocks."

"What?"

"Geology. It was either that or bugs. *Ento*mology. Science requirement. I took one look at the arthropods and ran out of there screaming like a girl. Rocks are better. Lesser gross-out factor."

"The older the better, huh?" he said. His smile was scampy.

" 'You are wide of the mark in some of your assumptions,' " I said, quoting his term paper comment back at him. "Not just any senior rock will do. Or rocker. I have my standards."

"I can see that. You also have very good taste in teachers."

"And you have duct tape on your loafers."

My gay boyfriend, Billy, says I'll only date men on Social Security. I guess Dr. Ivy qualified. After Paris, after Bub, who else was I gonna date? One of those callow, barely lingual Frisbee players out on the mall? Some kid in a sweatshirt and jeans who thinks it's funny to tie a bandana around his dog's neck and own an anthology of bongs? I'd had a hard enough landing; the only thing mitigating my transatlantic culture shock and giving me any comfort at all was my Tank, which had the unfortunate quality of making me simultaneously homesick. France haunted me, maybe even ruined me, for living anywhere else. I was almost sorry I'd ever experienced it. It made America feel strange. Flat. Juvenile. Was I *déjà* that jaded? To paraphrase my beloved Joni Mitchell, I was a free woman in Paris, I felt unfettered and alive. . . .

How to prolong that feeling? That was the question.

"*Dr. Ivy's,* like, worn-in Ralph Lauren," Dana said.

We were outside after English class, back on our bench, back discussing our favorite subject. Dana didn't know but I'd already gone out with our favorite subject, to a bookstore-café near his rented town house—bookstores being Dr. Ivy's "ridiculous compulsion"—and then to his rented town house. Excluding the fact that it took me an hour to get there in my Civic, I'd had a nice time tooling around the sticks with him in his Midget once I arrived. (The problem with dating your teacher is the same as that with dating a married guy; you can never go out-out. "Dates" happen inside hotel rooms or your, or in this case, his, place. The bookstore-café was an anomaly; we only went there that once and Dr. Ivy was a nervous wreck the entire time.) My good ol' boy hunch had been right on the money, too; Dr. Ivy was indeed a Dixie native. But, as he stressed, from one of the original thirteen colonies. (This is why I began calling him Bubba—to myself, anyway. It seemed fitting, plus it reminded me of Bub, consequently dragging out the French connection.)

Dr. Ivy was a real good talker and kisser and his breath was sugary. After lovemaking he'd say, "Thine, am I" as naturally as

Barbara Lewis sang "Baby, I'm Yours." Okay, so maybe it was a little odd that he kept his L.L.Bean rubber moccasins on in bed "for better traction." And when he took off the rest of his clothes he didn't hang them in the closet or put them on a chair, but rather laid them on the carpet just *so*, neatly arranged like a triple-homicide crime scene, like those victims' bodies outlined in chalk, except here there were no bodies, just clothes in suspended action like paper dolls'.

Other than that, he seemed relatively normal. For a preppy.

"Dr. Ivy's, like, comfortable," Dana said cluelessly. "You know, the wide-wale sandy corduroy pants that match his hair."

"Or the cuffed khakis," I said, lighting a Parliament.

"The emerald green Izod shirts with the turned-up collars. Or the fuchsia Lacostes with the turned-up collars."

"Layered under the long-sleeved pale blue or pale pink button-down oxford cloth shirts with the sleeves rolled up."

"And without the sleeves rolled up if layered under the Norwegian pullover sweater. The navy one with the white checks."

"Yeah. Layered under the shabby tweed jacket," I said, sipping my TaB. "With the tan suede elbow patches. Can't forget those."

"I think it's very sexy, that whole look," Dana said. "Those rubber moccasins—"

"Oh yeah, those rubbers are *hot*," I said, hoping my openly sarcastic tone would throw her off. That's the thing about keeping secrets; outward transparency is essential. "And their chain-tread outer soles really turn me on. All that *traction*. A-ttraction to the traction."

"And no socks."

"Ever! What *is* that? That is the strangest thing! Fifty-five pounds of layers upstairs—and not a single sock in the basement.

Dear, these people do not wear socks. Period. Maybe they're, like, allergic to The Sock."

"Who knows?" Dana said, shaking her head and shrugging. "We're Jews."

"*Eee-mily Dickinson,*" Dr. Ivy said, opening a poetry book. He was sitting on his unmade bed, naked but for his eyeglasses and the famous traction rubbers. A half-consumed Miller Lite and a watch with a woolen Stewart Black tartan band (striped grosgrain for spring and summer; plaid wool for fall and winter) were on his nightstand; three empty TaBs, a pack of Parliaments, an ashtray containing three Parliament butts, a lighter, one pair of diamond studs, my Sacré Cœur pendant, and one Tank were on mine. "Ah, sweet Emily," Dr. Ivy continued. "She was such a strange, heavy person."

"You're saying you think she was humorless?" I said. I was in my bra and panties, pulling my hair back and up into a ponytail in Dr. Ivy's bathroom mirror. It was late November, almost time for finals. I'd been sleeping with my professor for two whole months and, just like the fictional Harvard student in that *New Yorker* story I'd read, wondering how much longer this wonderfully inappropriate interlude would last. I believed that as long as I kept my Tank around, Dr. Ivy would stick around. He *had* asked me for coffee the first time I'd worn it, thereby taking us out of simple classroom flirtation and into, well, more explicit expressions of admiration. That meant my Tank had special powers.

"Emily's such an elusive figure and perplexing poet," Dr. Ivy said, "I just don't want to get trapped into saying any one thing. But man, the things that girl was thinking about up there in her room! She's good, but meshuga."

"Using My People's words on me, are we?" I said. I sat on the edge of the bed and put on my jewelry. I never wear jewelry to bed and I don't like other people to, either. There was this one guy I knew who refused to take off his African pendant. The entire time—the only time—he was on top of me I got smacked upside the head by the Cape of Good Hope.

"I like Yiddish," Dr. Ivy said. "It's colorful and expressive. Like you."

"Maybe Emily was meshuga because she needed a boyfriend," I said. "Or a girlfriend. She died a virgin, right?"

"There are no virgins left," Dr. Ivy said. "What are you doing?"

"I have to go," I said, reaching for my sweater. It and my jeans and socks were strewn across the end of the bed. I'd bought the sweater in France. It was a fitted black long-sleeved hip-length thickly-knit wool and cotton cardigan with, instead of buttons, an offset and very bold brass zipper that optionally snaked all the way up into a high turtleneck. I loved it so much I bought the same one in ivory. Now where'd I leave my black suede boots? (Is there anything better than black suede boots? Or, for that matter, black suede anything?).

"Don't go," Dr. Ivy said.

"I have finals I have to study for. Including yours, by the way. And seeing as you don't grade on a curve—"

"It's your curves turning me into a walking appetite," he said. "You create a desire in me to be unleashed. To erupt."

"I'm glad, dear, I really am. But I have to go now. It's a school night and I have rocks in the morning." I checked my Tank. Midnight. Jesus.

"Screw rocks. Stay. Come on. Sleep over. You know you want to. I have popcorn and there's a *Kojak* marathon on TV in an hour."

"Oh there's good salesmanship," I said.

"You know, I am very anti-intellectual in that sense," he said, removing his glasses and taking me in his arms. "Kojak has as much ego as he has little hair."

"I really, really should be going," I mumbled. He kissed me and I closed my eyes and gave in to the moment. *Quel* pushover. Dr. Ivy placed a pillow under where my hips would go, took off my lingerie, turned me on my belly over the pillow, and covered my back with his compact husky body.

"You really, really should *not* be going," he said. He was hard as a rock. And I knew my rocks. I was way past Rocks 101. "Since I've met you I've become really excited. Savoring words more, too. Valor loves a challenge. Let's get down to the heart of the matter."

"It's not my heart you're after, Dr. Ivy," I said, as he pierced another organ. "Is it." He didn't answer me. "Is it?"

The dance began. It was a slam dance. And a slam dunk. It felt good. I started to take hold of a free pillow but before I could he delivered *le grand slam*—so deeply and intensely that the force thrust my left wrist against the headboard.

My Tank shattered. Into many, many, many little pieces. Shards of glass and tiny metal and plastic bits ran down my arm and armpit and under me and across the sheets and down in the space between the top of the mattress and the headboard.

Dr. Ivy didn't notice. He was delirious. For a preppy.

"OH GOD," I wailed. "OH GOD. OHGODOHGODOH-GOD!"

Naturally, being a man, he had no idea what I was oh God-ing about. I held up my wrist behind me to show him. All that remained on it was an empty yellow "gold" frame and a black "alligator" strap.

"Huh?" he said. "What the—oh shit." He rolled off me and uselessly began trying to pick up the broken pieces while I sobbed for my destroyed Tank. "Oh my God, I am so sorry. I'll replace it. Or, or maybe we can fix it."

"You can't replace it, you insensitive watch-crushing pig!" I screamed, suddenly sounding fabulously French. "I brought this watch across an entire OCEAN! THOUSANDS OF MILES! My lover Bub GAVE me that watch! How're you ever gonna fix that? Huh? How?"

And you know what he said?

"You had a lover named Bub?"

The Just-Perfect Jean Jacket

Do you have a lot of jeans? I don't. Not because I'm into minimalism or self-restraint or because I lack closet space. I probably have more closet space than you do. And even if I didn't, I'm a woman. I can *always* find space in my bulging closets to add something new that I must have and cannot live without. This is partly because I don't have to share my closet space, thank God, and mostly because my best friend, Billy, changed my closets and therefore my life last year when he gave me a box of Huggable Hangers for Hanukkah. Huggable Hangers are the best invention since TaB. Really. They're that superlative. Created by Joy Mangano, an absolutely adorable Long Island inventor, divorcée, mom, and teeny-tiny brunette. Billy saw Joy and her hangers on the Home Shopping Network (HSN). That's where most of his disposable income ends up.

When I first opened my "60-pack Hanger Set," I didn't get it. They're hangers. They're attractive—I liked the velvety, ivory-colored fabric called "linen" they were covered in, and their elegant

gold hooks. But so what? I already owned—and Billy knew this—
at least 755,000 Container Store hangers, and everybody knows
they're the best:

- White plastic tubular hangers (29¢ each; a relative
  bargain, considering they're indestructible and look
  great)

- No-Slip Trouser & Skirt Clamps ($3.49 each; pricey,
  true, but you always have more dresses and tops than
  trousers and skirts—maybe)

- Grippy hangers ($6.99 for three; also pricey, but perfect
  for silk, satin, rayon, and any other fabric that slides off
  regular hangers. They're steel and covered with the same
  foam used on motorcycle and bicycle handlebars)

- Brass hangers ($3.99 each; I only use them in the coat
  closet and they look extra-nice when you pull them out
  for guests' overcoats and wraps)

Billy said, "Aren't those Huggable Hangers FABULOUS?
They're the only present I've ever given my mother that she didn't
instantly return!"

That gave me pause. Billy's mother, Louise, is one picky Italian. From Queens. That type doesn't mess around. So I decided to
give my Huggable Hangers a chance. I went to my guest room/
home office, site of the sole walk-in closet and my spring and summer wardrobe. I performed the ultimate sartorial acid test: I tried
a *very* slippery Isaac Mizrahi silk camisole in a soft off-white with
tic tac-sized lime green polka dots and an empire bust on a Huggable "shirt" hanger (the kind without a horizontal bottom bar). I
waited for it to fall to the floor. Unless it's on a Grippy, it always

falls to the floor. But it didn't. I jiggled the hanger. Nothing happened. Not only did my cami not slide off, it *wouldn't* slide off.

Wow.

Then I began doing what I automatically do: I pried my packed clothes apart on the closet rod to make room to hang something, anything, else up. But the Huggable Hanger was so slim—1/4" wide, and with an extremely narrow hook not much wider than a (gasp!) wire hanger's—that it just slid right in. I've never not pried! My heart jumped. I felt on the verge of Closetary Epiphany. I ran to my bedroom, which has a non-walk-in closet full of fall and winter clothes. It's not a big closet. When I moved into the apartment I immediately bought two Container Store closet rod doublers ($9.99 each). They're chrome-plated steel hooks with hardwood bars. You place them over the existing rod and they double your hanging space. Brilliant. Doublers aside, it is one jammed little closet. I wrenched out a pair of very chic Banana Republic black wool men's-style cigarette trousers hung on one of the old No-Slip Trouser & Skirt Clamps. I unclamped the trousers, folded them neatly lengthwise, and centered them on a Huggable "suit" hanger. Same result as with the Isaac cami. My pants didn't move *and* they slid smoothly back into the overstuffed closet. Joy Mangano! I think you may be an Italianate genius! The re-Renaissance has come! To New Jersey!

But wait. Hooold it right there, Michelangela Mangano. I have a skirt issue. I own a *lot* of skirts. I ran back to the spring and summer closet, where most of the skirts are. How're you gonna Huggably hang a skirt? I dug through the HSN gift box and there was my answer. That Michelangela thinks of everything! There was a twelve-pack of plastic Finger Clips that matched the hangers' "linen" color. The clips snap on to the bottom bar of the "suit" hangers and glide across (but not recklessly) to accommodate any skirts' waists.

I tried out a pair on my favorite one. It's from Anthropologie. I call it The Lettuce Skirt. It's unique. This handmade cotton skirt is a veritable work of leafy green wearable art. It looks like cascading, scalloped, rickrack tiers of crinkly cut-up iceberg lettuce bunches that descendingly evolve into romaine.

The new hanger plus Finger Clips hugged it perfectly and I hugged Billy.

"Aren't those Huggable Hangers FABULOUS?" I asked, squeezing him. "They've changed my whole life! I never knew it could BE like this!"

"Oh Chicklet, please," he said. "I'M the one who's changed your life." Billy always calls me Chicklet. "Chicka-boom, chicka-boom, Chicka-boom-boom-let."

While Billy stayed glued to HSN for last-minute Christmas presents—I'm the only significant Semite in his life—I spent the long Christmas week alone in my walk-in closet. This is one of the key advantages of being a Jew during the Yuletide. You can get an uninterrupted, guilt-free head start on your spring cleaning. So. Out with the old thick plastics, in with the new thin Huggables. You know that expression "It's gonna get a lot worse before it gets better"? By the time I finished changing the hangers on fifty-nine spring and summer tops, skirts, and pants, the inside of my closet and immediate environs looked as if a trunk had exploded. But! The lineup of clothes on the rod had magically shrunk. Not numerically—God forbid I get rid of any clothes—but spatially. I'd doubled, and maybe even tripled, my closet space. It was incredible. No bulk. These new hangers were miraculous. Now I could shop more 'cause I'd just created so much extra space!

I heaped the discarded fifty-nine tubular No-Slip Trouser & Skirt Clamps and Grippy hangers in extra-large plastic garbage bags, tied them up, and hauled them out of the way. I was exhausted. I poured myself a fresh TaB on the rocks, lit a Parliament, and studied the situation. I'd need at LEAST another 754,941 Huggable shirt and suit hangers *plus* Finger Clips for the rest of my wardrobe. (Although the Huggables weigh nothing and are plenty strong enough to hold even the heaviest winter coat—or even a little pink raincoat—I resolved to keep my coat closet brass hangers. They look pretty in there and they chime like glockenspiels.)

I went to *www.hsn.com*. Sticker shock: sixty hangers for $74.95 or 120 hangers for $149.95. That's almost $1.25 per hanger! Oh, and get this: $19.95 for each pack of twenty Finger Clips. Joy Mangano, what are you *doing* to me? It's a closeted conspiracy. No wonder you're so rich and I'm so not. I had to start small. Rome and my new hangers weren't built in a day. I ordered 120 "linen" hangers, typed in my much-abused Mr. Visa number and expiration date, hit ENTER, closed my eyes to brace for impact, and . . . the card went through. I was so excited and relieved that I ordered three more sets of 120 "linen" hangers, five twenty-packs of "linen" Finger Clips, and opted for Super Express delivery, which would put my Huggables in my hands in forty-eight huggable hours for an additional (and most *un*huggable) $31.45. My doorman Louie, who's a great guy and who'd have to sign for all those incoming heavy boxes and help me bring them up to my apartment, would be the perfect beneficiary of the rejected Container Store hangers.

But about jeans. I own some. But I almost never wear them. I love them on other people. But on me, no. They're uncomfortable. They're cold in the winter and hot in the summer and the material doesn't move, even if it's got a bit of stretch in it. Blue jeans are just too . . . definite. They're only good for one thing, really, and that thing is sex. Jeans are sexy, that's what they are. Now if you're definite about sex, then you should definitely wear them. I'm pretty definite about sex, but I like having looser, less definite options. It's like yoga. And yoga pants. And what you can do in them. Less inflexibility, more Downward-Facing Dog. You know what I'm saying, girls.

What I *am* absolutely definite about and always in the mood for is a jean jacket. I *love* jean jackets (hung on nice brass hangers). I love everything about jean jackets. Their inherent, effortless, seasonless, cuffs-rolled-back casual style. How they can cool out an otherwise potentially stiff outfit. How you can layer under and over them. How you can turn them into pillows and blankets on planes. How they don't wrinkle, even if you wear them insouciantly draped over your shoulders or with the sleeves wrapped around your waist. How cute they are over practically anything, be it bikinis, Bermudas or backless black dresses. I have a few of those.

I have nine different jean jackets. I thought I was fine with eight but I wasn't. These are the ones I had when I met The Romantic Russian Redhead, organized by hue:

Three almost-black midnight blues:

- The GAP's 67 percent cotton–33 percent polyester jacket. Hits at mid-hip, and has a regular collar, flat silver buttons down the front and on each cuff, and no pockets. (We hate no pockets.) Cut like a formal suit blazer you could wear to a business meeting. It's elegant but I'm a little afraid of it. I'm not sure why.

- The GAP's slouchy maternity version (really) in 100 percent cotton with a collarless banded neck, silver snaps instead of buttons, a just-below-the-derriere length, slashed side pockets (yay!), and stylishly exaggerated four-inch-high cuffs with three vertically aligned silver snaps on each. Perfect for those feeling-fat days.

🎀 Isaac Mizrahi for *Tarjay*'s totally cute, totally fun, fitted, and lightweight 100 percent cotton jacket, cropped to just below the adjustable (with two horizontally placed flat brushed silver buttons) waist, with flat brushed silver buttons down the front, bright orange topstitching, and wonderfully frayed seams (collar, front, hem, and cuffs) whose threads resemble faux fur. I LOVELOVELOVE this one. But Isaac, why no pockets?

The medium-to-powder blues:

🎀 H&M's LOGG (Label Of Graded Goods) brand, in thin 100 percent cotton, slightly bleached, in the basic V-neck, collarless jean jacket cut. Cropped to the waist, sans pockets (hey, it's H&M; I paid, like, $20 for it, what can you expect?), and with flat silver buttons and beigey topstitching. Good for poolside. This is the jean jacket you can get chlorine and your Banana Boat SPF on and not worry about.

🎀 LizWear's 97 percent cotton-3 percent spandex, tiny, very cropped, very fitted jacket, with a V-neck, neon yellow topstitching, buttons that look like old pennies, and slashed side pockets. Very cute and cozy. Extremely similar to the one Jessica Lange wore so sexily with her miniskirts and long naked legs and high heels in *Crimes of the Heart*. As Pauline Kael wrote of Jessica's jacket, "I came to like her faded denim better and better; she wears it like a badge of honor."

- ET VOUS's chunky, soft, I-want-to-curl-up-in-you 100 percent cotton and altogether amazing jean jacket. The blue is not too dark, not too light. The denim is not too heavy, not too flimsy. This is the French catwalk translation of the American cowgirl classic, cut like a dreamy cocoon (so you can bundle a lot of layers underneath) and with the coolest details—balloony sleeves, seven-inch-high twin flap pockets on the chest, deep side pockets (*bien sûr*), and flat silver buttons embossed with teeny five-point stars. How Ann Taylor ever got hold of this ET VOUS item I'll never know. It's one of my oldest jackets and far and away my favorite. I've worn it so much over the years that I recently had to have the entire thing reinforced and restitched from seam to shining seam.

The summery ice white:

- Gotta have it: the GAP's 95 percent cotton-2 percent Lycra spandex jacket with a collar, shiny silver buttons, white-on-white stitching, button flap pockets on the chest, side welt pockets, and a mid-hip-length hem. Perfect.

The vibrant lemon yellow duster (aka walking coat):

- Okay, technically this one's not a jacket-jacket. But it is styled like one, it is denim, and it is outerwear. And it's too cute not to mention. In 100 percent cotton by Cotton

Colors, knee-length, with a collar, silver snap buttons, four patch pockets (two on the chest, two on the hips), slashed side pockets (we love it), and a back vent.

I wore that duster to pick up pork and pignolis at Whole Foods. It was spring and my closets had finally been wholly Huggabled. They were so orderly and uniform, so tidy and sleek, so much more beautiful than any store display anywhere—that they just had to be seen to be believed. I decided to throw a little closet-warming Cuban dinner party with high-class calories and a "Let's Stay in the Closet(s)" theme for Billy and my best Jersey Girl friend Eman.

My menu: *puerco asado* (roasted pork tenderloin), *frijoles negros con arroz blanco* (black beans with white rice), *plátanos maduros fritos* (fried ripe plantains), guava-stuffed crescent rolls (served with dinner as the bread accompaniment), and *flan de coco con fresas frescas* (coconut caramel custard topped with fresh strawberries). Although the traditional Cuban salad is *ensalada de aguacate* (thinly sliced avocado with thinly sliced white onion and finely chopped fresh parsley, salt and pepper, and a simple Spanish olive oil/freshly squeezed lemon juice vinaigrette), I created another salad. It's not Cuban at all—*¡sacrilegio!*—but it is sensational and if I don't make it, Billy and Eman get cranky. The culinary equivalent of The Lettuce Skirt, it was inspired by the basic house salad at Capital Creations, a gourmet pizza place in, of all places, Raleigh, North Carolina. I used to live there, and this salad with balsamic vinaigrette is truly unbelievable. The pizza is great, too, but that salad—oh my God. I had this exceedingly cheap boyfriend at the time and he'd squawk about paying $8.20 for a double portion of, as he so charmingly put it, "fucking overpriced weeds." That was nice. So I analyzed the ingredients and came up with my own adaptation, which I've since enjoyed much more than I ever did that tacky old boyfriend.

for the
SALAD

Mesclun greens (about one handful per person, three if you're having Billy and Eman over)

Thinly sliced small red onion

Plain goat cheese (enough to incorporate a taste in each bite)

Toasted pignolis (pine nuts baked at 350° for about five minutes; use lots and be careful as they burn easily)

Garlic-butter croutons

for the
ORGASMIC VINAIGRETTE

(makes a half cup; you can double or triple the recipe):

2 tablespoons balsamic vinegar

3 tablespoons vegetable oil

3 tablespoons olive oil

¾ teaspoon salt

Freshly ground black pepper

1 garlic clove, pressed

Combine all the ingredients in a blender. Chill several hours or overnight. The dressing should be thick and dippy, almost like clotted cream. DON'T use a food processor or a hand whisk. The former will crush it into runny airlessness and the latter will make your arm tired.

✦ ✦ ✦

$\mathcal{S}$o *I'm* at Whole Foods, floored by the fact that their flaw-less boneless pork tenderloin (of which I needed five pounds—Billy alone eats for three people) and bulk pignolis cost the same: $10.99 a pound. I'd bought everything else at the regular supermarket. I love Whole Foods—who doesn't?—but I shudder to think of what this dinner would have cost had I bought all the ingredients there. I was happy, however, because it was a beautiful day and I was in my vibrant yellow duster, a white tank top, black pull-on flared terrycloth capris with vertical white stripes on the outer seams, black patent leather flip-flops with lean leather soles, diamond stud earrings, black sunglasses, red-red-red lips, and my fave huge black nylon tote from Banana.

As he rang me up, the cute redheaded cashier with a bushy red mustache said my duster's color reminded him of the Coldplay tune *Yellow*: "Look at the stars, look how they shine for you and everything you do, yeah they were all yellow."

"Is that a Russian accent singing this British song?" I asked.

"*Da,*" he said. "You're good."

"You stand out," I said. It wasn't the accent. This is New Jersey. *Every*body has an accent. What was different was that he was the only cashier who wasn't tattooed, multiply pierced, bald-shaven, ponytailed, braided, or dreadlocked.

"And you, Miss Yellow. You can call me Smirnoff."

"Are you a vodka?" I asked, laughing.

"I could be, for you. Then you would drink me, get drunk on me."

Yowza. Is this how you meet people? Standing on line at Whole Foods, about to buy a $55 piece of pig to celebrate the fact that you got new hangers? Smirnoff was pretty cute. He looked to be about my age, maybe a couple of years older. And he's from the former USSR. How about a British accent singing a song about the USSR? *Let me hear your balalaikas ringing out/Come and keep your comrade warm.* Hmmm. I was born in Cuba. Perhaps we *could* effect some cross-cultural intoxication.

"I'm Gigi," I said.

"Really?" he said.

"Yeah."

"We had a little dog in Moscow named Gigi. So cute."

"Yeah. Everybody's had a little dog named Gigi."

"True. But not in Moscow. Would you like to get a Jamba Juice with me? I have a break soon. A few minutes only."

"Oh, well, I would," I said, handing him my credit card and deliberately not looking at the total. "But I've got pork. I need to get it home."

"You come back?"

"I live all the way in Hackensack. I don't come out here that much."

"Come back," he said, looking at me tenderly. "Gigi." It was lovely, the way he pronounced my name.

"Okay," I said. "I will."

"When?"

"I don't know."

"Tomorrow?"

"No, I can't tomorrow," I said. "Next week."

"Okay. We go for a coffee. I live near here. With my parents and sister. We're in United States not even one month. Political asylum. New Jersey Reform temple sponsored us."

"Really? 'Save Soviet Jewry'?"

"*Da.* And you would save me."

"Save you from what?"

"World without yellow."

✦ ✦ ✦

My "*Let's* Stay in the Closet(s)" party was a hit. Billy and Eman oo'd and ah'd over my Huggabled closets and Cuban cooking and un-Cuban salad. (I never mentioned Smirnoff during dinner, afraid Billy might try to steal him.) Afterward, as I was cleaning up, I saw the Whole Foods receipt on the kitchen counter. The sum was $23.73. How was that possible? Smirnoff had charged me for ASSRTD MRCHDS, not for pork tenderloin. That sneaky little Russki. Was he immoral? Was this, like, a black market Moscow thing? Was he Russian Jersey mafia? Could this be true love?

The following week I went back to Whole Foods. There was Smirnoff, laughing and chatting up customers in his untucked green plaid poplin Lacoste shirt with the sleeves rolled up, Levi's 501 straight-leg jeans in inkiest indigo, and navy Sperry Top-Sider canvas espadrilles ringed in jute. Very cute. Only a certain type of man can get away with espadrilles. It's the same thing with fisherman sandals. If the man's straight—and I gathered that despite this one's distinct warmth and sensitivity, he was—he can't be American. I walked to the end of Smirnoff's long line, comprised only of women. Had I noticed Smirnoff's full lips before? They were like Angelina Jolie's. I bet they'd feel amazing. Anywhere.

"Gigi!" Smirnoff said. "You're back!"

"I have come back for the Jamba Juice," I said.

"In green skirt."

"It's not *green*. It's lettuce." Yes, I'd whipped out The Lettuce Skirt. I wore it with the pale blue LOGG denim jacket; a white tank top; a lacy white bra; Frye camel leather ankle-strap sandals with crisscross uppers, raw-cut edges, and leather-wrapped platform soles with 3" sculpted heels; a 20"-long gold and "emerald" grapes cluster pendant whose sparkly citrine stones perfectly matched the skirt's iceberg portion (I have matching pierced earrings but they're so heavy I can't wear them); white mother-of-pearl vintage cat-eye sunglasses with dark green lenses (which matched the skirt's romaine portion); my all-time fave Helen Kaminski natural raffia bucket bag; and Chanel-ified red lips. I know. Devastating. Thank you. Or should I say, *cnacuбo* (spah-SEE-bah).

"I'll be a few minutes only," Smirnoff said. "We meet outside."

Another beautiful spring day. There were dozens of pots of lilies in rows in front of the store. Their fragrance was heady. Smirnoff emerged with two paper cups of fruity smoothies. We sat on a wooden bench in the warm sunshine, clinked cups, and reached for our cigarettes at the same time. He smoked Parliament 100s. Nobody smokes Parliament 100s. Except us.

"It's funny, you know," I said, taking his proffered Parliament. He lit it for me and lit his own. "I feel like I know you somehow." I sipped my pineapple-strawberry-banana beverage. It was tasty with the cigarette. I'm all about health.

"*Da,*" he said, exhaling a long cloud of smoke. "I feel like this, too. When I saw you the last week."

"You did?"

"*Da.* I was studying to become psychologist at Moscow State University. Where Chekhov, Kandinsky, and Gorbachev studied."

"What about Dostoevsky? Did he go there?"

"*Nyet.*"

"Tolstoy?"

"*Nyet.*"

"Chagall?"

"*Nyet.* You know Russian artists," he said. "And clothes. I love what you wear." How uplifting. A heterosexual male who actually notices my outfits. A shaft of late afternoon sunlight streaked across Smirnoff's face, illuminating the depth of his green eyes, his ginger-colored eyebrows and eyelashes, and a few depressed acne scars across an otherwise smooth and very pale complexion.

"I know everything, dear," I said. "Except what I don't."

"I planned to marry girl I knew from there a long time, but it didn't work out," he said. "I went to work in my father's business."

"Which is?"

"Oh, several things. Properties more and more. So far, I'm never married."

"Me neither."

"The Russian women . . . not so good for me. All they talk is money-money-money. They want for American men to marry and come to United States. Only my mother and my sister are the good Russian women. The Jewish. Like you?"

"Yeah, I'm a member of the tribe," I acknowledged.

He laughed and touched my arm. The sun felt hotter. I got up and removed my jacket. "I am in love with that blue jean," he said, pointing to my jacket. "May I try?"

"Try my *jacket*?"

"Mmm," he said, standing up. "Why not?"

"Um, I don't know if it'll fit you," I said uncertainly, handing it to him. He put out his Parliament and put on my jacket. Smirnoff was well proportioned and not too big, maybe 5'6" and 140 pounds. I'm a lot shorter and lighter than that but my jacket *almost* kind of fit him.

"Hi, you!" said a dreadlocked Whole Foods clerk to Smirnoff. She was heading to her car. "Is that your sister?"

"*Nyet*, it's my Gigi," he said.

"Oh. You guys look like twins," she said. "See you. Cool jacket."

"You didn't charge me for the pork," I told him.

He smiled and said, "That's good, no?"

"Well, you know. No. I mean, I am a freelance magazine writer. It's always feast or famine. So I appreciate it. But—"

"We have dinner tonight, yes?" he said, pulling off my jacket. "Russians have a grand love for writers. You must know. It's our history. You like crêpes?"

"I thought you were from the steppes, not the crêpes," I said.

He laughed and said, "Crêpes are like blinis from home. But . . . not blinis. Not home." He looked a little choked up. I guess one man's madeleine is another man's buckwheat pancake.

I was right about those lips. I got to touch them later that night with my fingertips and then with my lips after the crêpes. Smirnoff and I fell in together as though we were reuniting lovers from a past life; he seemed so familiar to me. Maybe it was the Russian connection—my maternal grandfather, Boris, was Russian—or maybe it was the way we instantly got one another. Being in (and occasionally out of) bed with him felt natural, terrific. Smirnoff loved taking long bubble baths together and having three-hour-long conversations about anything. He noticed everything and complimented me constantly on my clothes, my hair, my accessories, even my makeup. At night he'd pat lip balm on both of our lips and massage my hands with lotion and blow on them. It was like having an older sister and a nightly slumber party, only we weren't related and he had a penis.

"I think the ghosts of the poor dead Romanovs sent him over

here, dear," I told my girlfriend Eman one weekend while Smirnoff was at work. Though Billy still hadn't met him, Eman had and they'd hit it off. They both have Russian-born fathers. Smirnoff told Eman she was a dead ringer for Sofia Coppola. I said Sofia should be so lucky.

"He's adorable," Eman said, exhaling her Marlboro Light and pouring half a can of TaB and half a can of Diet Pepsi in a glass full of chopped ice and a straw. We got into this habit initially as an experiment to make the TaB go farther when they raised its price. Then we just loved the combo. Diet Pepsi mellows and deepens TaB's sprightly, hollower flavor. "So? Should we start looking for wedding favors?"

"I love him," I said, pulling a platter of "craps" out of the fridge. I'd invented this variation on crêpes to appease Smirnoff's hunger pangs for his beloved blinis. What is it with Russians? They're so unhappy and dour. It's always winter in their souls. Consider Garbo in *Ninotchka*: "I'm so happy, I'm so happy! Nobody can be so happy without being punished." Russians hate it in Russia and they come over here and then once they're here all they can do is talk about how much they miss Russia but they never go back. Even my animated, occasionally cartoony boyfriend suffered bouts of that distinctly inert, self-pitying Siberian darkness. Smirnoff would play Barbra Streisand singing "The Way We Were" over and over and over, getting smashed on his alcoholic namesake and weeping inconsolably for the lost motherland. I'd feed him a few craps to perk him up, figuring the sugar couldn't hurt. He'd devour them, smoke a Parliament, start feeling better—and go right back to the vodka-distilled misty watercolor memories. I'd remind him that Tolstoy, who was a million laughs, said, "If you want to be happy, be." In other words, happiness is a decision. Smirnoff would gaze at me with these big deep sad wa-

tery faraway eyes—and resume his inconsolable weeping for
Mother Russia.

But. That's no reflection on my craps. Not at all. They went
over so well, in fact, that I actually went out and purchased a $35
thermostatic temperature-controlled, nonstick, 7½" cooking sur-
face, electric crêpe maker. It's an excellent but strictly one-trick-
pony appliance. Now *that* is craps commitment. (Alternatively, you
can use an 8" skillet. It's not as much fun.)

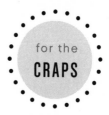

for the
CRAPS

(makes 18)

1 cup sifted all-purpose flour

2 tablespoons sugar

3 large eggs

¼ teaspoon salt

¼ cup whipping cream

1 tablespoon butter, melted

1½ tablespoons cognac

1 cup whole milk

Whisk flour and sugar with eggs; add salt. Thin mixture
with cream. Add melted butter and cognac. Mix well and
dilute with milk to make a very thin but smooth batter;
strain through a sieve. (The batter may be kept overnight in
the refrigerator.) If you have a crêpe maker like mine, just
follow the manual's instructions. Otherwise, over medium-
high heat, brush a skillet with a little melted butter. Add

two tablespoons of batter and tilt the pan to coat the bottom. As soon as the crêpe begins to brown (thirty to forty-five seconds), use a spatula to lift and flip the crêpe over to finish cooking on the other side, about twenty seconds. Stack each crêpe between paper towels or they'll stick to each other and then you'll spend hours peeling them apart.

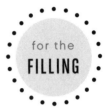

for the
FILLING

These are all approximate proportions; just be generous with them because more really is more in this recipe; and preheat the oven to 350°F.

1 16-ounce container sour cream
2 sticks sweet (unsalted) butter, melted
3 cups white granulated sugar

Line a large baking pan (either 13" x 9" or 18" x 13", depending on the number of craps) with aluminum foil. Pour enough melted butter across it to coat it evenly, including the edges. Now here's the drill for each crap: slather sour cream across it in a medium layer, almost to the outer edges; pour about one tablespoon of melted butter on the sour cream; sprinkle enough sugar over that (about a tablespoon) to "hold" the melted butter; roll the crap into a cylinder; line it up on the pan; pour melted butter over it end to end; and sprinkle sugar over all of it. Repeat until every crap is prepared. Bake for about thirty minutes. The craps are done when they're golden brown and caramelly. Allow to cool and harden for one hour. They're wonderful served cold or at room temperature.

"These are, like, one of the best things you've ever made, EVER," Eman said, delicately holding a crap in her fingers and tearing into it like a Russian mama bear. "Jesus, they're good. Do we think he deserves 'em?"

"Well, he inspired 'em," I said. "And his sex is inspirational when he's not being suicidal. And he's always bringing home, like, six-dollar lobsters and three-dollar filet mignons."

"Some of those melancholy ones can be so romantic and emotional," Eman said optimistically. "Think of the great Russian pianists and composers. Rubinstein, Mussorgsky, Tchaikovsky, Rachmaninoff, Stravinsky. They're all depressives. It's like they're playing with no skin."

"Smirnoff's circumcised," I said. "It's not that. There are other problems. Well, one."

"They all have problems," Eman said, reaching for her third crap. "Each girl just attracts a different kind of loser. Is he, like, the kind of loser you could see yourself married to?"

"I'm afraid he'd try to wear my wedding dress," I muttered.

"Excuse me?"

"He's into my clothes."

"He likes what you wear?"

"He likes *wearing* what I wear. Like, the jean jackets. There's not a single one he hasn't worn. Like, around the house. He just throws it on. It's like the opposite of 'the boyfriend sweater.' "

"Even the shrunken Isaac one from *Tarjay?*" Eman shrieked. "With the fringe?" The girl knows every single thing in my closets. She has a photographic memory and she's getting almost everything in the will. So it behooves her to know.

"He loves the Isaac especially. He says the orange topstitching matches his hair."

"Oh my God."

"But he only wears it around the house. It'd be too gay outside."

"You're sharing *your* clothes with your boyfriend."

"It's not my plan! The other day I found him puttering around the apartment in the ET VOUS jacket—"

"Oh *no*, not the ET VOUS!"

"Yes! With my navy blue T-shirt and the madras walking shorts!"

"The cute blue plaid ones with the elasticized waist?"

"They were like hot pants on him. It's a nightmare. Thank God his feet are too big for my shoes!"

"Thank God is right. Can you imagine him trying to squeeze into the Coach metallic pewter sandals you maxed out your poor little Visa on? The ones *I'm* going to inherit and wear every day for the rest of my life with bare legs or black tights?"

"I love men and I love my clothes," I said. "I just don't love men *in* my clothes."

"Sounds like tranny time to me."

"*Granny* time?"

"*Tranny*. Hello! *Transvestite*."

"WHAT? He's NOT! You're crazy."

"You said you think the Romanovs' ghosts sent him over here? In Anastasia's dressing gown, maybe. Cross-dresser. Did you know that Anastasia used to wear her father's underwear?"

"You're OD-ing on craps."

"*Tran*-syl-va-ni-aaa!"

"Stop it," I said, laughing. "You're freaking me OUT."

"Hey, at least we know he's straight. It's better than him being some gay drag queen in campy outfits or a transsexual, God forbid. I know, find yourself a super-femme jean jacket. Something even more femme than the Isaac with the fringe. And then see what he does."

"That is psychotic! You're saying I have to go shopping to find something to *repel* my lover instead of attract him? I've NEVER done that! That goes against everything I've spent my whole life—"

"Could I get some of these craps to go?"

Of the 131 stores I hit at Garden State Plaza, only Nordstrom had the just-perfect jean jacket. Of course. Nordstrom is fabulous. (I always feel sorry that New York City people don't have one.) I first had to narrow it down to two different denim jackets, both sweet and less than $100, in stonewashed blue cotton: one was a relaxed, hip-length Cynthia Max with no closure or buttons; flap pockets on the chest (but no side pockets); vented sleeves and double rear vents; and little ruffles trimming the collar, placket, front hem, and cuffs. The other one, by Rafaella, was fitted, cropped, and traditionally styled, with logo'd and weathered gold buttons, roomy slashed side pockets (yay!), darted seams in front and back, and a black velvet collar.

I loved them both. Ruffles? Too cute! Black velvet? So great with faded denim—but you couldn't wear it half the year. On the other hand, I like pockets. Hence, the black velvet won. Smirnoff would *never* wear black velvet, especially if I pinned a brooch on the bodice. I had the perfect one in mind, too. It's the size and shape of an ostrich egg, all antique gold filigree (which would pick up the gold of the jacket's buttons) with rhinestones and a rectangular centered jet (which would pick up the black of the jacket's collar). I envisioned holding the new jacket aloft, shaking it at my boyfriend, and shouting, "Smirnoff! Here! I DARE you to wear THIS!"

I tiptoed into the bedroom and there he was. Standing in front of my full-length floor mirror. In my Hanro shell-pink mercerized cotton high-cut briefs with the wide lace waistband. And nothing else. Head cocked, he was hugging my Huggably hung Lettuce Skirt over his trunk. Of all the skirts in all the towns in all the world, he picks my Lettuce.

I don't think so.

"Okay, buddy," I said. "Now you've gone too far. Surrender the skirt."

"I only try it," he said. "It's sooo beautiful."

"Hand me The Lettuce Skirt. Thank you. Now hand me the Hanros. Thank you. Now step away from The Lettuce and the panties. Back away slowly and no one gets hurt."

"What's in Nordstrom bag?" he said excitedly. "Something new?"

"Yeah," I said with a sigh. "I bought you a new jean jacket."

9

Black Silk Slingbacks

The Cheffy Chef was one kick-ass Big Daddy. He embodied Tennessee Williams's big, brash, manly, room-displacing Southern patriarch with gusto, guts, and great charisma. He approached his cooking the same way. The full-time chef, part-time actor, and nonstop charmer hailed from the Pelican State, where boys know their alligator pears from their muffulettas long before they ever sprout facial hair. At fiftyish, my Cheffy Chef, as I liked to call him, had a shaved head, a tiny ruby in his right ear, and a tattoo of Bugs Bunny where the sun don't shine.

It was the Bunny that got me. *Le lapin.* The tattoo, or "tat," was an homage to The Chef's favorite Merrie Melodies character and to his favorite dish, Rabbit Tenderloin Étouffée with Smoked Andouille Sausage and Caramelized Onions over Aromatic Rice. That was the first food he ever made for me on our first official date, at my apartment. I'd never dated a chef before and I didn't think I'd be into wabbit, wascally or otherwise. But when I tentatively put the first forkful in my mouth, I nearly passed out from the pleasure. It was just indescribably delicious. I greedily scraped and lapped up every last drop on my plate as though it were my last repast on Earth. I was like one of those little lab mice that press the bar for more coke until they die. I was like my cat Lilly who's so addicted to kitty crack, er, treeeats, that I'm looking into some serious rehab for her. I was like Scarlett O'Hara on her New Orleans honeymoon with Rhett Butler: "If you don't stop being such a glutton," Rhett told his insatiable new bride, "you'll be as fat as the Cuban ladies and then I shall divorce you."

Being a Cuban-born lady myself, I really could've lived without that remark. I got Rhett's drift, though. And you know what? I didn't care. I didn't *care*! I ate and swooned and ate and swooned. The Cheffy Chef regarded me all the while with his chin in his fist and knowing bedroom eyes whose expression made me feel undressed even with all my clothes on. Not that I had that many on: black lace Banana Republic bra and matching bikini panties; an Asian-flavored, flowy, light as a breeze, semisheer Banana scooped U-necked ankle-length double-layered sleeveless silk slip dress with 8" slits on either side of the hem, in black overlaid with ivory plum blossoms; delicate brushed pewter drop earrings whose shape echoed the dress's blossoms; and, if you count these as "clothes"—and God knows I do—COCO perfume, red coral lips, and matching toenails. If you're gonna do black—The Cheffy Chef's favorite color—you MUST MUST MUST brighten it up with red extremities.

Speaking of extremities, The Cheffy Chef and I were extreme—and pure and unquenchable—appetite. It was chemical and caloric. What could be more desirable than delectable food, sensually prepared and experienced? I've always loved people from Louisiana, chefs especially. When I was just a little kid, I enjoyed watching that Cajun cook with the big belly and suspenders,

Justin Wilson, on PBS. He'd say, "How y'all are?" and "I garon-
tee!" in this corny patois and cook up vast tubs of mouthwatering,
fattening goodies like chicken and sausage jambalaya and pra-
lines. Through the years I've met and written about a few of these
chefs. There was one in North Carolina several years ago who,
like the so-called City That Care Forgot's Cheffy Chef, looked
uncannily like a younger, slightly slimmer cross between Burl
Ives and Jimmy Buffett. Is it genetic accident? Something in the
Louisiana water? Who knows? Beyond appearances, there are
definite qualities these guys all seem to share. They're lusty, fun,
rascally. A man who's big and comfortable in his own skin and can
woo you with a luscious Cajun stew of shellfish or chicken served
over rice—that is the man for me. In the case of The Cheffy Chef,
I was his fair game, a hare. And he was the hound. When finally
that first night I sighed, sated, and dropped back in my chair, he
reached his huge paw of a sautéing hand across the gulf of the
mixed fruit oilcloth I'd laid across the table, and clutched mine in
his. His skin felt rough; all that cooking, probably. But I liked it.

"Happy, *chère?*" he asked rhetorically.

"Delirious," I mumbled blissfully. Ahhh. I was so stuffed I
could barely breathe. I'd really made a Porky Pig of myself, eating
all that Bugs Bunny.

"More wine?" he asked, reaching for the Pinot Noir.

"Why not," I said. I lit a cigarette off one of the unscented
pillar candles I'd clustered on the table, and finished my crisp
vino while The Cheffy Chef put on coffee and looked through
my eclectic CD collection. He chose *The Glow* by Bonnie Raitt.
One of my favorite albums. Very underrated. The sassy opening
number, "I Thank You," was a palate-cleansing, head-clearing
tonic. *Now I know what the girls are talking about when they say that
you're . . . ooo . . . I wanna thank you, baby, thank you, honey.*

A man who's not obviously hunky can make you feel as though he is, and then some. The Cheffy Chef could do that to a woman. By his laid-back, unhurried, open-minded, good-natured, entertaining ways. By his hilarious dirty jokes and anecdotes about a couple of Cajun Country good ol' boys named Boudreaux and Thibodeaux who live down on the bayou. By his intelligent, earthy, sloe-eyed eroticism. The Cheffy Chef was smart and *alive*. He was a man I wanted to talk to while he cooked for me for the rest of my life. Around him, I swirled, swelled, and seesawed with desire. With him, I felt no less warm, yielding, or sweet than the caramelized onions in his Rabbit Étouffée or the Champagne-soaked raisins in his *riz au lait*, rice pudding. I wanted to eat with him, talk with him, go to bed with him, and stay there forever. With chicory coffee and beignets and deep kisses in the morning and then more coffee and beignets and steamy lovemaking. At least until lunch.

"Oh no, ma'am, I'm taking care of *you* tonight," The Cheffy Chef said as I started getting up to clear the table for coffee and dessert. "You sit down and enjoy the music."

Who WAS this guy? Don't you love him? They don't make 'em like this north of the Mason-Dixon Line. By then Bonnie was singing, *I feel a glow . . . Here I am . . . What I need is a friend. What I need is someone . . .*

If food be the music of love, eat on! We polished off a chilled bowl of pudding. We drank steaming black coffee and talked and talked and talked while I smoked and smoked and smoked. As the candles flickered, he regaled me with true adventure stories about growing up in Breaux Bridge, in St. Martin Parish, home of the annual crawfish festival. The Chef's papaw (grandfather) and daddy used to take him out hunting and fishing in the low swamps. What a magical marshland, with mysterious Spanish moss dripping and shady bald cypress trees and brown pelicans hoarsely crying and diving for fish in black water. Cheffy learned to catch catfish (flathead, aka mudcats and channel and blue), sacs-a-lait (aka white crappies, aka white perch), and of course crawfish. He shot ducks (mallards, canvasbacks, blue- and green-winged teals). He shot alligators. He shot turtles. He shot rabbits in his own backyard at night and helped his grandmama make Rabbit Sauce Piquant or "outstanding" Alligator-Turtle Sauce Piquant with 'em. He told me a little about his late wife who'd died in a car wreck near Lake Charles a long time ago, and a lot about his obsession with his new vocation and oldest love after Southern literature, cooking.

I'd never known anyone like him. He dazzled me.

The Cheffy Chef extended his hand, pulled me to my bare feet, and escorted me to the living room for a slow dance to the CD's last song. My favorite one. The gorgeous, quiet, aching, soulful, and very deep "(Goin') Wild for You Baby." It wipes me out every time I hear it: *Nights I can't sleep, my tears are cheap, I'm losin' hold of my senses . . . It's makin' me wild for you baby. . . .*

We kept dancing, or rhythmically moving together, anyway, long after the CD ended. Except for one last sputtering candle, the room was in darkness. We gradually stopped swaying, our arms still entwined around each other. He whispered, "You turn me on," and kissed me. Soft. Hard. Soft. Really hard. He did everything just right. I hadn't been kissed so expertly in . . . years. His kiss liquefied me. The Cheffy Chef was direct yet meandering, languorous yet urgent. Masculine. Passionate. Hungry for me. And that was how, on a very tender midnight in spring, a Cajun-Creole with a secret leporine tat became my great new Big Easy lover.

That's the thing about dating an intellectual who's also a chef: You really have to love a man who reads and cooks dissolutely and will do so for you. My Cajun-Creole was a rising mid-Atlantic star, a word-of-mouth local legend. He had a master's and a Ph.D. in American Southern Literature, and a breathtaking command of Louisiana cuisine. He'd left teaching Walker Percy and Flannery O'Connor for teaching Bananas Foster and roux after a distant relation died and left him a small fortune. With that he'd recently opened a little café-market-cooking school not too far from my place.

The first time I walked in I smelled Tabasco with a trace of . . . sautéing onions and celery and bell peppers, maybe? The so-called trinity of Louisiana cuisine. Through the speakers Hank Williams Jr. sang about jambalaya, crawfish pie, and something called *filé*—not *filet*—gumbo. There were native foodstuffs everywhere from hundreds of Louisiana vendors, including a freezer full of alligator andouille sausage.

It was like being in a foreign country.

"Bonjour!" said The Cheffy Chef, emerging from the back of the store. *"Bienvenue!"* He was wiping his hands on a small towel with tiny green alligators on it, and was wearing a blue apron with a single gigantic red crawfish on it. The crawfish had a man's head with a *toque blanche*. Its caption read, *"Laissez les bon temps rouler!"*

"Like those shoes," The Chef said in a seductively raspy Southern drawl, shaking my hand. "They're darn sexy."

"Thanks," I said. "By the way, I'm Gigi. And these are my Minnie Mouse shoes."

"They'd be amazin' in black, too, *chère*," he said, still holding my hand. "I like anything black. Or blackened."

I was wearing my beautiful Ann Taylor corally pink red suede pumps called Azalea with rounded toes and stacked 3½" heels. They were so adorable, so impractical and surprisingly comfortable for a heel that I'd bought the same pair in Blue

Heron, a brilliant, deep, bright turquoise blue. What is it with (as my girlfriend Sharyn and I call it) Ann Tay-Tay? They go through these fleeting colorful starburst candy spurts of utter grooviness, where you want to buy everything in the store, especially the shoes and jewelry. But then, I don't know, something happens and they revert to a forgettable, blah, buttoned-up, government-working-stiff beige. Get with it, Ann Tay-Tay! Have some TaB Energy! Have some fun! Wake UP! Show women it's OKAY to wear a color besides light brown, medium brown, and dark brown! Especially in spring. Seriously. I just got the new Talbots catalog and—DON'T LAUGH—they had 3,000 cuter 'tudes in happy, juicy colors like honeydew, wild watermelon, sweet lemon peel, and saffron orange. You just wanna EAT them.

The reason I was in such pretty shoes was because earlier that day I'd been to see one of my magazine editors for a breakfast meeting about a freelance piece he wanted me to do, so I had to doll up a bit. I wore my pumps with thirteen different Chanel lipsticks (the number it took to replicate the exact shade of the shoes); a starched black and white vertically striped men's-style shirt in cotton and spandex by Isaac Mizrahi for *Tarjay*, with roll-back cuffs in black with white polka dots (too cute!); a black just-above-the-knee linen/rayon Ann Taylor pencil skirt with a rear kick pleat; my Ann Taylor teardrop white pearl and silver earrings; my Fossil watch with a faux black alligator band and square silver and white face; a heavy silver DKNY chain-link bracelet with a dangling heart tag (very similar to the classic one from Tiffany); two white seed pearl necklaces and bracelets that resembled magnified grains of rice, from the American Museum of Natural History's fantabulous pearl exhibit's gift shop; and my gigantic floral fabric-lined GAP tote in metallic silver. This is about as conservative and officey as I get.

"This is some place you got here, Chef," I said. "I love it. It smells . . . all spicy."

"And here I thought I was the one who specialized in hot stuff," he said, looking me down and up and down again. "Ya hungry?"

"I could eat."

"I was just foolin' aroun' with somethin' for tonight when ya strayed in here like a little cat in hot pink Minnie Mouse shoes. We're only open for dinner so far and even that's just part-part-time." He pronounced *time* like "tahm," slow and musical. I love Southern accents, I think they're beautiful. "Come have a seat. Come on." He led me around the dozen café tables and chairs to the open demonstration kitchen, with its wraparound bar and stools. "Ya like a beer? I got some Abita in here. Or maybe a glass of wine? Got a nice Pinot Blanc chilled, too."

"Wine is good," I said, taking it all in. He popped a new CD into the player and handed me the case. Irma Thomas's *Live! Simply The Best.* "I've never heard of her."

"Shame on you. Irma's major. The illuminati of R&B all know her. She's the soul queen of New Orleans."

"I thought the queen of soul was Aretha. Or, is Aretha."

"'Retha just had a better press agent," he said, handing me a glass of cold white wine. "I saw Irma perform for the first time when I was a teenager. I snuck into a black nightclub in New Orleans. She was the name act. She was very pregnant, as I recall, just a couple years older than I was. She's sung duets with Bonnie Raitt. Saw 'em on TV on New Year's Eve once."

"Really? Bonnie? I love Bonnie!"

"All right, now hush. Enjoy your wine and listen to Irma while I make your lunch."

Irma said, "Now we gonna get to—'It's Raining,' you said,

huh baby? Let's drip!" And she began singing the smoky, sultry, sexy R & B song: *It's raining so hard, Looks like it's gonna rain all night, And this is the time, I'd love to be holding you tight . . . , But I guess I'll just go crazy tonight.*

As things began heating up in the kitchen, I lit a Parliament and sipped the Pinot Blanc. It was light and fruity. I studied The Chef's displayed collection of bottled hot sauces:

- Blair's Sudden Death ("This Is It")

- Blair's Death ("Feel the Heat . . . Feel Alive")

- Blair's After Death ("The Tough Guy")

- Red Hot Mama ("Desire the Fire")

- Mean Devil Woman ("Born on a bayou years ago was a Mean Devil Woman. She was so hot the 'gators ran, the swamps boiled, and the hearts of men caught fire at the very sight of her. Cajun radar has shoved the very essence of this woman in a bottle to be rubbed on whatever you dare . . .")

- Calido's Batch #114 ("Pain Is Good")

- Calido's Batch #37 ("There is a point where pleasure and pain intersect. A doorway to a new dimension of sensual euphoria. Where fire both burns and soothes. Where heat engulfs every neuron within you. Once the line is crossed, once the bottle is opened, once it touches your lips, there is no going back.")

- Hot Sauce from Hell ("Devil's Revenge")

- Café Louisiane Hotter'n Hell ("Shake The Devil Out Of It!")

- Smack My Ass and Call Me Sally ("Big Chet was a bad dude. The kinda guy that would steal the wooden leg from a handicapped person. The kinda guy that would hide the dentures from his grandmother. So it was no surprise when his wife, Janice, slipped some of this homemade hot sauce into Chet's moonshine. After one sip, Big Chet fell to his knees and with a tear in his eye shouted, 'Well, SMACK MY ASS AND CALL ME SALLY!' ")

"Are you kidding?" I said, picking up the Smack My Ass and examining its label. "The boy has his britches down and his cheeks are all red. Is this real?"

"Oh yes, ma'am. Smack My Ass."

"Oh my God, please tell me you're not using this. I mean, it's not even noon."

"Those crazy ones are just for fun, mostly. For Chicken Sauce Piquant—that's what we're havin'—it's Crystal." He handed me a bottle of Crystal Cayenne Garlic Sauce. "It's vinegar based. Baumer Foods makes it. They've been around since 1923. Started out makin' jams and jellies. They've got a magnificent art deco neon signboard on Interstate 10-West. It's a New Orleans icon."

"I've only been to New Orleans once. In 2001. I was my girl-friend Gwen's maid of honor at her wedding in City Park near the flying horses."

"The flying horses, ah yes."

"Yeah. The painted ponies. The carousel. All her family's from New Orleans. Where is that signpost?"

"Just past downtown on the right-hand side as you're goin' west toward Baton Rouge and the airport. It's a fellow dressed as a chef, mixing a great big vat of strawberry preserves."

I'd overlooked a hot sauce: Katrina Storm Sauce ("Category 5 flavor and Category 3 heat. Guaranteed to spice up your life with a flood of flavor!").

"Oh," I said. "This is . . . gee. I don't know what to say about this one."

"I was down in New Orleans last Christmas and picked it up," The Chef said, taking a sip of his Abita beer. "Katrina. What a fuckin' heartbreaker. None of my kin were impacted but I have a lot of friends who . . . Anyway, I go back and forth a lot. To help out as best as I can. I get homesick real easy. Still not sure I wanna stay up here forever. So, Baumer Foods, the sign I was tellin' you about? Their factory got hurt pretty bad in the hurricane. They're relocating to Reserve in St. John the Baptist Parish. On the east bank of the Mississippi, northwest of New Orleans."

I heard a sudden boom of thunder outside, a big ripping crack, and suddenly a sunny blue sky day turned into a rainstorm.

Drip drop . . . It brings back memories of the time that you were here with me . . . Counting every drop, About to lose my top . . .

The Chef served up the Chicken Sauce Piquant over bowls of steaming white Creole Rose aromatic rice, and pulled up a stool next to mine.

"*Bon appétit, chère,*" he said, clinking his beer bottle with my wine goblet.

"*Merci,*" I said. I took a taste. Have mercy. "Oh. My. God. This is . . . Oh. My. God. This is sublime! It has a little bite. But a good kind of little bite."

"Well, some like it hot, as they say."

I slapped him on the upper arm. "This is orgasmic!" I shrieked.

"Happy to oblige," he said, smiling a smile that was giving me *incredibly* inappropriate ideas. I wanted to throw this Cajun-Creole

buck on the floor and hop on him with my cottontail like a wild
Cuban bunny. But not before I finished my Poulet Piquant. It was
too good: fried chicken simmered in chicken broth and smothered
in a tomato-based sauce incorporating the holy trinity (onions,
celery, and bell peppers), with fresh tomatoes, garlic, hot peppers,
cayenne pepper, and other glorious springs of pain.

"You're killing me," I said. "You know what my lunch usually
is? Two Quaker Crunchy granola snack bars, one Nature Valley
banana nut or peanut butter bar, thirty-seven TaBs—sometimes
with Diet Pepsis mixed in—and thirteen Parliaments. All con-
sumed at my computer while I'm on deadline. That's all I do. My
whole life: one big deadline."

"That ain't good," he said. "Person's gotta live a little. Have
some fun."

"I know! Smack my ass and call me boring! But I'm in the
middle of writing a book *and* freelancing. I have no life."

"Really? Well, I gotta hear all about that. Over some coffee
and sweets, maybe. Can't let the 'Storyville' lie fallow."

"Say what?"

"Storyville's New Orleans' old red-light district. This dessert's
so sinful that I couldn't think of a better name for it."

I could. Rough sex. Try this on: Chocolate-chocolate
chip-toasted black walnut pound cake bathed in a butter-Bakers
chocolate-heavy cream-maraschino cherries-Grand Marnier-
Courvoisier sauce. In case that wasn't *quite* slutty enough, The
Chef plunked a wanton scoop of vanilla bean ice cream on the
side.

"I can't have that whole slab!" I said. "I'll have to be hospital-
ized."

"Eat what ya like, sugar. I have to-go plates." I regarded him
quizzically. "Doggie bags." I smiled at him and dug in. Oh. Oh.
Ohhh.

"I'd smack your ass and call you Sally," I said. "But I'm about to . . . well."

"About to what?"

I felt myself blushing. Must've been the cherries. My face felt hot and the rest of my body felt flushed, as though I was turning the color of my Minnie Mouse shoes.

"Could I have a glass of ice water?" I said.

"Hell, you can have a whole bottle. Got some Evian in the fridge." I nodded vigorously. "You okay?"

"Oh yeah. I'm just—thanks." I glugged the cold water right from the bottle and dabbed my mouth lightly with the back of my hand. My skirt's waist was strangling me. I wanted to tear off all my clothes and get *real* horizontal. What I needed was a nap. What I needed was . . .

It's raining so hard, Looks like it's gonna rain all night . . . I wish the rain would hurry up and end, my dear.

"I should go," I said, feeling just the opposite. "This has been, uh, well. A sincere sensation. It was wonderful. You're very . . . endowed. Oh God, what I meant is talented. Gifted. You know what I'm saying."

"I'm gettin' ya."

"Thank you so much for an amazing lunch. I think that may be the best lunch I ever had—outside of the Delta."

"It was my pleasure," The Cheffy Chef said, smiling. "When you think I'm gonna get to see you again, *chère*? Scootin' outta here like a rabbit in the rain."

"Well," I said, looking out the window and taking off my shoes, "in a week? That's how long it'll take me to digest this lunch. Chef, do you by any chance have, like, a plastic to-go bag or something I could put these in? They're suede and it's still pouring out there."

"I believe I can satisfy your needs, ma'am," he said. "Let me wrap up some Chicken Sauce Piquant and Storyville for ya, too. Wouldn't want the starving artist goin' hungry on me. 'Sides, ya never know when ya might get the urge for *piquant* again."

The Cheffy Chef was in the midst of gutting and renovating a big, stunning loft with high ceilings and rows of windows overlooking nothing but sky in a landmark ten-story turn-of-the-last-century warehouse with a freight elevator like the one in which Glenn Close and Michael Douglas gymnastically trysted in *Fatal Attraction*. The loft was urban and gritty, historic and romantic, with unbelievable potential. But. The Cheffy Chef's a *guy*. Hence, his domestic priorities—kitchen excepted—are wrong. Let us begin with the concept known as a girl-friendly bathroom. There was no shower in the bathroom, only a tiny (48") white antique bathtub, which, while charming (ball and claw feet), had no nozzle attachment. That is fine—for a baldie. But not for a person of my coiffure. And where would all my coiffure care products go? Because let me tell you, after washing and conditioning my hair, it takes me a good twenty minutes to de-frizz, blow-dry with diffuser attachment, color protect, curl enhance and individuate, root lift and thicken, separate all into piece-y-ness, hold it still, and shine it. Naturally, this involves a battery o' product: TIGI Protein Protective Spray, KMS Hair Play Molding Paste, KMS Curl Up Curling Balm, Rusk BloFoam Root Lifter, TIGI Catwalk Texturizing Pomade, Sebastian Shaper Hair Spray Styling Mist for Hold and Control, and Rusk Shining Sheen and Movement Myst. I know it sounds like a lot of product, and I'm not saying it isn't. But if I skip any of those steps I could pass for Don King's Caucasian twin. And it wasn't just the hair products I was wondering where to put. Where would the eye products—my contact lens case, contact lens cleaner, and contact lens conditioner—go? Or the skincare and body products? Or the five billion makeup products?

No, it was better all around to be at my place, with the girl-friendly shelf and counter spaces and two separate bathrooms. Initially, I was a little worried about introducing a new father figure to Lilly, my cat. She's used to having a stay-at-home-single mommy doting on her all day and sleeping with her and only her all night in the fabulous Tempur-Pedic. My previous mattress had been dying a long and hideous death for years. I think my parents had bought it for me for my first apartment—so we're talking really ancient. Like most of my ex-boyfriends, but less lumpy. Once I'd given The Cheffy Chef's predecessor the heave-ho several months before we'd met, I'd decided to symbolically heave that old mattress ho, too. Kind of like Farrah Fawcett in *The Burning Bed*, only not quite as *Movie of the Week* spectacular. I did, however, spectacularly splurge on a Tempur-Pedic and asked my gay boyfriend Billy to come break it in with me. We laid down on it side by side.

"Isn't it wonderful?" I asked. "It, like, holds you. You feel weightless. Like you're in a cloud."

"I don't get it," he said. "It's not comfortable."

"You can't appreciate anything good or Swedish! That's what this Sleep System is, you know. It's a Swedish sleep revolution from the Land of the Midnight Sun. That's what the cute guy at Brookstone told me when I tried it out at the mall. One of us should be dating him."

"There is something seriously wrong with you."

"The Tempur-Pedic's made of high-tech, NASA-engineered, ergonomic foam with memory cells, so it dips and conforms to your body," I said, parroting the cute salesman's Sleep System speak. "See how it leaves an impression where you are and then fills itself back up when you're off it? See how I can roll around on my side but your side stays perfectly still so I'm not disturbing you? See how it's ideal for sleepless people with back problems?"

"Yeah, ideal for sleepless RICH people with back problems."

"It feels heavenly."

"It feels bizarre," Billy said. "Like we're sinking into the Baltic. You said it's Swedish, right? Hahaha."

"We're not sinking into the sea, silly," I said. "We're relieving pressure points."

"Thank you, Patty Hearst. You've drunk the Kool-Aid. Stockholm syndrome."

"The Sleep System comes with a twenty-year warranty and they threw in a complementary—and complimentary—Swedish neck pillow. See? It's anatomically correct—don't even say it—for your spine. Look at the pillow. Put your head on it. The dual-lobe design supports the curve created by your head, neck, and shoulders to properly align your spine and give you a perfect night's sleep!"

"Zzz."

Billy's the exception to every gay rule. A *normal* gay guy would LOVE the Tempur-Pedic. Heterosexual female kitties, however, are another story. Lilly was so freaked out over losing "her" old mattress *and* the way the memory cells depressed under her tiny paws that she boycotted the new bed and me for days. Then, as felines are wont to do, she changed her mind. Then she could not be moved from it. She and the bed have since become attached like Velcro. See? Those sleep-savvy Swedes do rule! Cajun-Creoles, too. My Cheffy Chef took to my Tempur-Pedic like a 'gator to a swamp.

As for Lilly, it was love at first sight with The Chef, who

scooped her up, cradled her in his arms like a baby, and called her "my sweet little darlin' " until she purred herself into an orgasmic stupor. Later, in a gild-the-Lilly move, he fed her a morsel of his signature creation, based on a famous dish from the original Mulate's, a wonderful Cajun restaurant in Breaux Bridge: deep-fried catfish nuggets over Creole Rose rice, smothered in crawfish étouffée.

It was right up there with the treeeats.

At least that's what Lilly purred.

"*He's adorable!*" Gwen whispered. It was late afternoon and we'd arrived early at The Chef's store for a cooking class, which he held several times a month. Gwen, a D.C. financial analyst friend, had come for a weekend visit and was eager to check out my new "gentleman caller," as she called him, of several months, and her fellow Louisianan.

"Yeah," I whispered back. "Watch these women here tonight. They'll fight each other to have his baby before we get to the bread pudding with Wild Turkey sauce."

"Hey you," The Cheffy Chef said, hugging and kissing me proprietarily on the lips. "And you must be Miz Gwen. You're gorgeous! You look like Alicia Keys but better. Hell, Halle Berry best watch out!"

"You're a real flirt," Gwen said. Was she blushing? "So nice to meet you, finally. I've heard a lot about you."

"Likewise," he said. "Gigi told me all about your weddin' by the flyin' horses. Still married?"

"Barely," she said.

"Too bad, too bad," he said. "All the best ones are barely married. 'Cept my gal. And I'm not lettin' her get away."

Gwen raised her eyebrows and smiled at me.

"You two look so cool and pretty in your summer dresses and sandals," he said. Then, calling his two high school interns: "Ashley? Maggie? Would y'all be sweethearts and get these ladies some—you drink wine, Miz Gwen?—some nice cold Pinot Blanc? I got to get set up. But first, would you like to see my pots?"

"Sure," she said. At this point he could've asked her to set her hair on fire and she would've looked for a lighter.

"Ohhh!" Gwen said. "You have the right pots!"

"Yes, indeed," he said proudly. "Only Magnalite will do. My earliest memory is of my Grandmama teaching me to cook red beans and rice in these. She taught me all the Cajun and Creole basics and then I just sorta took it from there, improvisin' and tastin' for color, texture, and layers of flavor—my personal holy trinity. When you're poor you got to get inventive and draw on what's on hand. We had shrimp, oysters, crawfish—"

Gwen sniffled. She covered her mouth with her hand. She blinked away a falling tear. I suddenly found myself in the Dixie Gothic Twilight Zone. Why were Magnalite and shellfish making Gwen cry? Ashley and Maggie served us our wine and began setting the bistro tables with fresh-cut magnolias in little opaque milky green Depression-era Jadite glass vases. Gwen and I walked into the kitchen and clinked glasses with Cheffy as Nancy Wilson began crooning "How Glad I Am" over the speakers.

My love has no bottom, my love has no top . . . I can't stop loving yooou . . . and you don't know hooow glad I am.

My boyfriend opened cabinet after wooden cabinet, revealing all manner of seasoned, thick, cast-aluminum cookware. I'd seen him use those saucepans, skillets, frypans, Dutch ovens, and casseroles with their thick black knobs and handles a thousand times. I never gave them a second thought.

"Ohhh, you have them aaall," Gwen said. "Now I know you know what you're doing 'cause you have Magnalite pots."

"Heh-heh-heh," he said. "That's right, baby. They're at least fifty years old. Got 'em from my Grandmama, all passed down to me."

"That's how it happens," Gwen said, welling up. "They last forever."

"What is going ON?" I said. "Please tell me. I hate being the only Yankee."

"Let me tell you how important these pots are," Gwen said, accepting The Chef's proffered handkerchief and collecting herself as he got his roux under way. "People *will* these pots to their children. It's very deep. No respectable Southern kitchen goes without Magnalite. It's a staple. You must, you must, you must have it."

"Oh," I said.

"My mom lost all her pots in the flood," Gwen said. "Her mom had given them to her when she married my dad."

The café went quiet, except for the soft whirr of ceiling fans and Nancy Wilson's voice.

My love has no walls on either side, that makes my love wider than wide, I'm in the middle and I can't hide loving yooou.

The Cheffy Chef took his pan of roux off the flame and put his big hand over Gwen's. He looked at her tenderly.

"Oh Gwen," I said, embracing her. "I'm so sorry. You never told me this."

"Mom and my sister, they evacuated to Atlanta right after you called me," Gwen said. "Remember, you called me at work and you said, 'You have to get your family out of there.' You'd been watching it on TV."

"I remember," I said.

"But Daddy, he wanted to stay to help evacuate more people," Gwen said. "And you know what happened to him."

"What?" The Cheffy Chef said.

"He wound up alone on the roof of their house in Gentilly," I said. "They were fucked."

"They were super-fucked," Gwen said. "Because they're near Lake Pontchartrain. They didn't have a chance! They're on Wellington Avenue right between the two levees that broke, the 17th Street Canal and the London Avenue Canal. I'm picturing my dad sitting on the roof of his house, in his straw cowboy hat, a short-sleeved silk shirt in a pastel Easter print, Bermuda khaki shorts, and Top-Siders—"

"A black man in Top-Siders," I said.

"I know," Gwen said. "And there was my tour bus driver dad, clutching his Cuban cigars, bottled water, medicine—he's a diabetic—and a bag full of toiletries and clean underwear, screaming, 'Hey guvnuh Blanco! Forget about finding me some housing. Whacha gonna do about my wife's Magnalite pots?' "

"He waited three hours to get rescued by a fire department boat," I told The Chef.

"They wouldn't let him bring the Magnalites," Gwen said. "There was no way *to* bring them. Then when they finally got to go back to the house, everything'd been steeped in sewage. Pots were floating around in that septic soup, in mold. Fifty different colors of mold! It's disgusting."

"Magnalite's not porous," The Chef said. "It's salvageable. You could boil it with some bleach, it should be okay."

"Well, it wasn't destroyed," Gwen said, sipping her Pinot Blanc. "But it was soaked in sewage and all that other stuff floating around and the mold took over when the water receded. It was pretty nasty. Mom didn't want to risk it or take any chances. She toyed with the idea of cleaning them but you can't. Where the handle meets the pot? You never know what's in there and if you can ever kill it. My brother-in-law bought her a brand-new set from a restaurant supply place. She uses it but it's not the same. The old stuff was the romance that is family, the generations of meals prepared using those pots. That's why it hurt my mother so much."

"They sell 'em at Wal-Mart," The Chef said. "I bought a roaster last Christmas at the one in New Orleans. But only 'cause I couldn't find the one I wanted on eBay. That's where you get the best Magnalite. Anyway, this new roaster? It's a whopper. Biggest one I've ever seen."

"Bigger than *you*?" I said playfully.

"Well, that ain't sayin' much," he replied.

"You exaggerate, honey," I said.

"Hey Miz Gwen, I had this girlfriend when I was a teenager," he said, changing the sad subject to cheer us all up. I knew what was coming. "We were makin' out one night by the bayou. It was our first time. She looks down at my Woody Woodpecker and says, 'Who're you gonna show a good time to with THAT little thing?' And I said, 'ME!' "

I'd heard that one before, but it still cracked me up whenever he told it.

Then, for his oldie-but-goodie coup de grâce: "So there's this kid and he wants to lose his virginity. Goes down to Storyville to get the job done by a professional. Lady takes him upstairs to her room and when he takes off his clothes, she busts out laughin'. His pecker's just *puny*. Even got the word *Shorty's* tattooed on the end of it. A minute or two later, the gals downstairs hear a woman's high-pitched screams comin' from that room. 'AIEEE!' They rush upstairs and open the door. 'OH MY GOD!' they holler. Boy's standin' there, with his pecker at full mast. It says, 'Shorty's Café and Truck Stop Highway 80 Going East Chattanooga Tennessee.' "

It was an intimate crowd that night, ten women and one guy. Locals, mostly. They'd each paid their $65 (not including drinkies, sales tax, or tips) to be entertained and learn how to make an appetizer, an entrée, and a dessert:

- Chicken & Andouille Gumbo—*Gumbo* is the West African word for "okra," and this is the ultra-classic, hearty Cajun roux-based stew served over aromatic pearly white rice. Yummy.

- Crawfish Étouffée—Words cannot convey how good this is. The summit of Cajun cooking (in my non-Cajun opinion), it combines crawfish tail meat, fat from the heads of freshly boiled crawfish (truly), the holy trinity, and a mess of other delicious things in a very rich, creamy sauce, served over Creole Rose rice.

- New Orleans Bread Pudding with Whiskey Sauce—I'm obviously biased since I was, you know, sleeping with its creator. But seriously, this was the best fucking bread pudding I've ever eaten (and I know my N'Awlunz desserts). I'm a fool for nuts, and Cheffy's scandalous recipe included walnuts *and* coconut in addition to the usual raisins. And that diabolically debauched whiskey sauce—*c'est si bon*. Butter, sugar, condensed milk, half-and-half, and Wild Turkey. When The Chef served Gwen a fat slice and poured warm sauce over it, she actually asked for MORE sauce. He lit up, said, "Bless you. Bless you, my child," and submerged her slice. The one guy in the class MOANED and banged the table with his forehead in rapture after tasting his portion. As for the women, well, they looked the way women look after they've had the best sex of their lives. I accompanied several outside for a "postcoital" smoke.

"Oh wow," Gwen murmured, fanning her flushed face with her hand. "That was intense. So sensual. Not just the food, it's, like, the way he prepares everything. He must blow your mind in bed."

My mind and my everything else. I smiled my best, most enigmatic Mona Lisa smile. Gwen was right, of course. The Chef exuded vitality and imagination in and out of the Sleep System. Later that night while we were in it, he held me and said, "Know what, *chère*? I feel kinda homesick. All that talk with Gwen 'bout Magnalite and New Orleans and—"

"Yeah, I can understand that," I said.

"Maybe I miss home more than I know."

"Do you?"

"I just suddenly feel . . . I don't know. A pull."

"It's natural," I said. "Urge for going back to the 'gators?"

"Well. All that Katrina talk. Hm."

"It's all right," I said, squeezing him and stroking his big bald Big Daddy head. "It's all right."

By autumn, The Cheffy Chef announced he wanted to "take a break and go home for a while." He couldn't define "a while." That summertime conversation with Gwen, his own rest-

lessness, and a free-floating wetlands withdrawal had really af-
fected him. Now, I could've pulled a Thelma Houston. I always do
when I'm in love: *Don't leave me this way, I can't survive, I can't stay
alive without your looove, oh baby please*, and so on. Very disco angst.
Instead, I somehow reined myself in and channeled Sting: *If you
love somebody, set them free.* Sting isn't a Louisianan, but he had the
right idea. Because knowing my Cheffy Chef as I did, I knew one
thing for sure: You can take the boy out of the swamp but you
can't take the swamp out of the boy. If this Big Daddy-Big Easy-
Bugs Bunny-Cajun-Creole was my destiny, me begging him to
stay up north or to take me with him down to be his bayou babe
wouldn't fast-forward it. I know that all sounds really mature but
maturity had nothing to do with it. The truth was, he had the
money to take as long a break from his business as he wanted, we
weren't living together or engaged, and I was on a book deadline.
As my beloved Leonard Cohen says, "For the writing of books,
you have to be in one place." It's amazing how quickly and cogently
your priorities click into place when you realize that. So. Men may
come and men may go, but I had a book to finish.

And a certain pair of shoes to find. I wanted to surprise my
Louisiana-bound beau by giving him a send-off he'd never forget.
He'd inspired it, actually. He'd once confided he had a fantasy
about the two of us in his café powder room for an intermezzo, as
it were, while oblivious patrons argued about who The Cheffy
Chef was more like—Paul Prudhomme or Emeril Lagasse (answer: neither)—and gorged themselves on bacon-fried shrimp
and cheese grits.

"And with you playing hostess and dressed in all black," he'd
said. "Especially with some killer CFM's [Come Fuck Me's, or
sexy high heels]." God, I love a Cajun-Creole with a dirty mind. (I
guess that's redundant.) Jack Nicholson must be at least partly

Cajun-Creole because I read a quote on *www.jack-nicholson.info/ news/jack-adviser-diane-keaton-to-be-more-sexy.html* where he "advised Diane Keaton to be more sexually adventurous if she's aiming for a chance to win his heart. Keaton revealed her feelings for Jack on the *Oprah Winfrey Show*. She said, 'He's [the] man I love. But he doesn't feel that way about me.' Nicholson responded later in the *New York Daily News*, 'Just like me, everybody loves Diane Keaton.' Afterwards he jokingly advised Keaton to 'invest more in heavy sexual acts. Not too distorted, but at least interesting in nature.' "

Café powder rooms and CFM shoes? Not too distorted at all, and certainly interesting. Interesting enough to make my man reconsider and stay? Well, probably not. But at least we could go out in style. With a bang. I knew just the kind of bang I needed, too. Black. *Peau de soie.* Strappy. Slingback. Open toe. Four-inch stiletto heels. Nothing articulates "Meet me in the bathroom before we get to the French Quarter praline sundaes" better than that.

I know what you're thinking. Because it's what I was thinking. This girl's on her way to *www.bergdorfgoodman.com* to blow hundreds of dollars she doesn't have on a new pair of CFM's she'll never wear again. But no. I did not. I did no such thing. I was being Sting-Zen-like, you recall, not Thelma Houston-Studio 54-like. Hence, no trendy desperation. Hence, I looked deeply and eternally within—my very own wardrobe. Yes I did. Waaay in the back, stacked on the top shelf in a dusty, five-year-old *yin* Ann Taylor shoe box, I found The *yang* Shoes. Black. *Peau de soie.* Strappy. Slingback. Open toe. Four-inch stiletto heels. Life is good. You have to let it flow from you. Then again, I live from book advance to book advance, so thank you Jesus, I do shop with the long term in mind. And because I go easy on my shoes, these

babies looked brand-new and timelessly fabu *and* The Cheffy
Chef had never seen me in them.

Now for the dress. Obviously, it had to be a dress; I was playing hostess-powder room Storyville tart and—*oh, my God. Hello!
I'd forgotten all about you! Honey, honey, honey, how are you? Long
time no wear. Let me look at you.* Isn't it sad when you talk to your
clothes? But this dress was worth talking to and talking about.
Also from Ann Tay-Tay, also never seen by The Chef, it was the
dress I wore to Gwen's wedding and it was PERFECT. *Comment
ironique!* Gwen had instructed all the women in her bridal party
to wear black. Black in New Orleans in May is like Blackglama in
Death Valley in any month. I wouldn't recommend it. But, as maid
of honor, your job is to support, agree with, and do everything the
bride says and wants. I'd gone with a fluttery, light, lined, silk
chiffon, sleeveless, pullover, tea-length sheath with a V neck and
back, and two long, sheer strips of silk ribbon behind the neck to
tie into a loose bow or knot, or not. Very beautiful in its utter sim-
plicity, very Audrey.

Accessories. This was a dress you could easily overdo them
with, if you gave in to the "Look how you can dress it up or down!"
attitude. There are times when overdoing is the only way to go, no
question. But getting, say, your Faraone Mennella bracelet's gold
chain links or toggle closure caught in your lover's nether regions
in a powder room while there are people outside waiting for dessert? Not so much. (That would be almost as bad as wearing a
jean jacket with jeans. Almost.) So. Sim-ple. Since I was on a
Gwen's wedding-all-black-New Orleans-Cheffy Chef-"I like anything black or blackened" roll, I chose exactly what I wore with
the rest of the original ensemble: a pair of really cool, really beautiful Stephen Dweck drop earrings with sterling silver hoops from
which dangled clusters of smooth and faceted black coral beads,

and a quasi-corresponding Meredith Frederick faceted black onyx and hematite beaded roll-on bracelet with tiny silvery beads appearing here and there. Both LONG-TERM pieces were from *www.bergdorfgoodman.com*. Why are you looking at me like that? Because I went there after all? Because I said I wouldn't? Excuse me, that was a while ago *and* I didn't go there for my SHOES. Nor did I go there for my DRESS. Thank you. I contained myself. Really. I did.

Life is too short to have sorrow. You may be here today and gone tomorrow. You might as well get what you want, so go on and live, baby go on and live, tell it like it is . . .

Aaron Neville's sweet tenor trilled as The Cheffy Chef chopped, stirred, mixed, and chatted up the private party of twenty-five that had reserved the café for the night. It was wonderful watching him in his element, in his apron, casting his Southern Louisiana spell, showing off in his self-deprecating way, being funny and charming, making others feel good and excited, letting *les bon temps rouler.* And not just that; he was a hell of a good teacher—patient, knowledgeable, calm, passionate about passing on his culture, culinary and otherwise.

And passionate about me with him in his powder room. Except for getting one of my black silk heels caught in the brushed nickel Pegasus toilet paper holder, it was poetry. As for the ensuing praline dessert, that too was poetry.

And as for The Cheffy Chef and me today, well, we'll always have I-M's and late-night phone calls. As he put it so well in a recent e-mail, "We never boiled over or degenerated into a messy food fight; it just sort of got eased onto the back burner, where it continues to simmer."

Gladiator

There is no couture at Back in Touch. This is not a high-fashion kind of place. For one thing, it's in Teaneck, New Jersey, where dressing up means wearing your best satin yarmulke. For another, the massage therapy shop is run by a guy who hasn't so much as TRIMMED his waist-long hair in a decade. The rainbow-hued woven anklet tied to his left ankle? That's been there since the Dead concert tour of '94. And his tie-dyed T-shirts, some of which date back to the Woodstock era, were tied and dyed by his wife, who used to have pink hair and teaches sewing at home.

This is Krunchy's world. The rest of us just get massaged in it. Krunchy has a real name—Allen—but considering what I've told you so far, don't you think my name for him is better? It really works with the peace, love, and granola-ness of it all.

My rapport with the Krunchmeister began in February of 2005, when I ran into my friend Eddie by the mailboxes. He and his partner, Michael, live in my Hackensack high-rise. Eddie remarked that I looked "kind of really tired, honey." That was a polite way of saying that I looked like shit. And he was right; I did. I'd spent the preceding two backbreaking years tethered to my computer, writing my first book. Becoming a published author had been my dream since I was a child. Now that it was upon me, the phrase "Be careful what you pray for, you just might get it" suddenly leapt to mind. Suffice it to say I was a lot cuter when the process began.

For the first few months of it, I was a model of functionality. I got up early, made my bed, took a shower, and fixed my hair every morning before eating a nourishing breakfast. Then I'd put on Actual Clothes, not fleece hoodies and cotton tank tops and yoga pants and Dr. Scholl's sandals/flip-flops/bare feet—my eventual, and ideal, writing uniform—and apply full makeup.

I was also very disciplined about my yoga and calisthenics, at least twice a week. I can't function without regular exercise and stretching. When I pretend I can, well, it's not pretty. If I let it go past two weeks, my friends start treating me like I'm having a psychotic PMS break, and wisely flee. I do my one-hour workout right in my living room. (There's a fancy "fitness center" in my building but the only person I allow to watch me contort myself into unholy positions is Lilly, my kitty.) All I need is my Champs Sports mat, a pair of old, soft, pink Danskin ballet slippers, and pair of lime green three-pound hand weights I bought at Nordstrom's Athletic Shop for Women. (Why would you use ugly weights if you can have adorable lime green ones?) My dad, a retired medical doctor, gave me an illustrated exercise booklet twenty years ago and I based my routine on that. Over the years I've added, subtracted, and improvised on it.

What a good girl, right?

It didn't last.

Even *I* thought I was being unbearably good. And so began the inevitable erosion and slide of my goodness:

- 🦋 Sleeping in (rationale: I worked sooo late last night, and listen to that rain coming down out there! Is there anything better than the sound of rain while you're all warm and comfy under the blankies?)

- 🦋 Not making the bed (rationale: It's not as if Lilly or I mind; we're in it most of the day anyway.)

- 🦋 Postponing the "morning" shower until after *Letterman* (rationale: I'm still clean. It's not as if writing all day takes major physical exertion.)

- 🦋 Not doing the hair (rationale: Hey, whatever. At least I washed it and worked in some Kiehl's Crème with Silk Groom afterward.)

- 🦋 Eating Dunkin' Donuts for breakfast (rationale: They have the world's best coffee. I'm not getting dressed and driving all the way out there just to buy one large, toasted almond, black. With these gas prices? That's ridiculous. I'll pick up a toasted coconut donut and six Munchkins while I'm at it. Make it worth my while.)

- 🦋 Sartorial decline (rationale: Like I'm really gonna run out to Dunkin' Donuts on Hackensack Avenue in my beautiful ivory Banana Republic wool and nylon cropped and lined jacket with the ragged silk edges; matching skirt; and costumey, Mia Farrow in *The Great Gatsby*, flapper-y, pearlized ivory leather Mary Janes with the three thin buttoned straps, pointed toes, and curved kitten heels! Please. It's weird enough that I actually wore it all to work at home one day! Haul out the hoodies, et al.)

> Not doing the makeup (rationale: There's this bumper sticker—"Today was a total waste of makeup." Well, my makeup takes an hour to apply. Really. An *hour*. I could be spending that same time reading *www.pagesix.com*, or, better yet, sleeping. And, when I absolutely MUST go out, like, to buy food I can't have delivered, I don chic black sunglasses the size of mainland China, and Chanel's Gladiator red lipstick, to camouflage my otherwise barefaced sins.)

> Yoga avoidance (rationale: It's not normal to "like" exercising, only *having* exercised. I can't do it in the morning because it makes me throw up. I can't do it mid-afternoon, because that's when I'm working. I can't do it after dinner, because that makes me throw up, too. That only leaves between 5 and 6 p.m. to do it. Oh my God, is it 9:30 already? Really? Damn. Did I miss *American Idol*?)

You can see why I not only *looked* "kind of really tired, honey," I *was*. My body—my back, especially—had calcified and turned to stone. It was like something out of the Bible.

"You know what you need?" Eddie said, pulling the new *Vanity Fair* out of his mailbox.

"A life?" I asked. "A new man? Several new men?"

"A massage. I couldn't LIVE without mine. I get one every week. It's unbelieeevable. You HAVE to have one."

"I love massage. I haven't had one in ages. Where do you go?"

"Right in Teaneck. Five minutes away. Off Cedar Lane on

Water Street. Past Bischoff's, on the right. It's called Back in Touch. So good."

I didn't know Back in Touch, which opened in 2001, before I moved to New Jersey. I knew Bischoff's, though. Everybody around these parts does. The sweet, old-fashioned "Ice Cream Parlour-Candy Store-Luncheonette-Soda Fountain" has been around since 1934 and still makes all its own to-die-for ice cream. Their servers wear these retro paper hats and bow ties—it's too cute. It's always packed with soccer parents and their kids, teens with iPods, and Orthodox Jews buying kosher ice cream cakes to go.

Hm. A little massage, a little ice cream, a little cake, a little Judaism. What could be bad?

Past the Empire Hunan restaurant and the Nail Garden salon, past the Dirty Dancing and Body Shapers studio ("Belly Dancing Classes Available NOW!"), across from Bergen Stamps Collectables and 1 Day Shirt Service and next door to the Chon-Ji Academy of Martial Arts and the Training Grounds Body Building & Boxing for Men & Women is Krunchy's shop. Its slogan: "All We Knead Is You." Perfect! I needed to be kneaded.

I don't know what Christie Brinkley looks like on massage day but I know what the rest of us look like. No little pink raincoats, no peach panties, no red ballet slippers, no mother-of-pearl earrings, no backless black dresses, no COCO perfume, no faux Cartier Tanks, no just-perfect jean jackets, no black silk slingbacks. None of that. We look—yuck—real. Me, I looked like a drowned, nearsighted, white-footed mouse with wet hair, glasses, a huge black down coat, and white leather Keds. The quintessential PLEASE IGNORE ME, JUST DRIVE BY AND SPLASH

MUD ON ME outfit. Hey, it wasn't as if I was going to meet a boyfriend and required my man-magnet props! After all, massage is the great equalizer, not the cotillion.

The exclusive concession I made to vanity and self-respect was, as always, Gladiator. Gladiator is like brushing my teeth. Indispensable. It's a Chanel lipstick. It's THE Chanel lipstick. The perfect, warm, bright, orangey-red that doesn't turn my pink lips blue. There's no living without it. Cosmetic companies don't seem to like making warm reds anymore. Go to any counter at Bloomies. All you'll find are cool blue reds—not that the average makeup salesperson can distinguish these. This is why I have to stockpile Gladiator. I do have some other warm reds, really old ones, that I use to make the Gladiator go farther. Some are a little dried out, but they haven't gone rancid: Max Factor's Pure Poppy. Prescriptives' Chinese Vermillion. Yves St. Laurent's #57. Estée Lauder's Parallel Red, Classic Red, and my all-time fave gloss, Mandarin Pop, of which I have only two left and they've long since discontinued. NARS's Flareup gloss. M·A·C's Redd lip pencil. Gladiator Chanel sisters Midnight Red, Paradise, Colère de Coco/Coco Crimson, Rouge Captif/Coco Peach, and Coco Mademoiselle gloss, which is shot through with tiny gold particles.

I keep all or most of these in my nylon LeSportsac Travel Cosmetic bag. I love my fat, double-zippered little bag. It's designed by fashion illustrator Sara Schwartz, a nice, slightly manic Jewish gal in New York. Her work is lively, funky, über-girly, cartoony, and kitschy. I used to own a simple plain black LeSportsac makeup bag. Big mistake. HUGE. NEVER use anything but the most conspicuously colorful kind. I learned this lesson the hard way when I lived in Raleigh, North Carolina. I was a features reporter there, and, being a massage aficionada, wrote a story about Darryl Foster. He's a fantastic masseur who worked in a medical building. He turned into my dear friend and my therapist. One day I went for a typically out-of-this-world massage. Afterward, as I blissfully floated down to the lobby, I couldn't find my car keys. I decided to empty my big black purse on the black carpet to search for the keys. Now, because I don't wear any makeup except lipstick to a massage, I had on my big black sunglasses. I found my keys, threw everything else back into my big black purse, and drove away. When I got home I couldn't find my makeup bag. If you're a woman, you know this is cause for alarm. I looked in the car. Nothing. I looked through my purse again and again. Not there. Oh *shit*. I must've left it on the carpet in the lobby! Because it was a *black* makeup bag, and my purse was *black* and that carpet was *black* and I was wearing *black* sunglasses indoors like one of the three blind mice—I lost my bag. Or rather, someone STOLE my bag and all its contents: the good Tweezerman tweezers. The only pretty contact lens case in the world. The retractable Clinique powder brush. The retractable Chanel lip brush. The eyeliner/lip pencil 2-in-1 Lancôme sharpener with a big hole and a small hole. The Chanel pressed powder. The 100-sheet pack of Shiseido Pureness Oil-Blotting Papers. The Kiehl's Lip Balm #1. The irreplaceable Clinique hand mirror that I got in one of their gifts-with-purchases. It's too painful to recount the lipstick and coordinating glosses. We won't go there. Way too tragic. This is why I think there should be makeup insurance.

The inside of Krunchy's shop is blue. A soft, pale, soothing blue. The light-filled space is clean and cozy, about 750 square feet, with a front desk and waiting area with some regular chairs and massage chairs, a back room, a bathroom, and three "session" rooms with dimmers.

That's it.

Though he spent the first decade of his therapeutic career at the nearby The Spa at Glenpointe, Krunchy's sole focus, like that of his seven Back in Touch colleagues, is on massage. There are no frills. There are no waterfalls or facials or mani-pedi's or hair removal or styling or fluffy terry cloth robes or paper sandals or candles or wedding makeup artists or fancy prices to match. (Sixty sublime minutes of Swedish massage cost $75. That's cheaper than gassing up your SUV.) You'll never read about Back in Touch in *www.dailycandy.com*. What a relief. I love Daily Candy but there are times when I just can't take one more hip and happening fashion/food and drink/beauty/arts and culture/fun/services/ travel item in my AOL inbox. It's exhausting.

Krunchy greeted me and I looked up to him. Literally. He's over six feet tall. And lanky, about 165 pounds. He's a young fifty-three, with a long, narrow face, round, gold-framed glasses, tiny studs and hoops in his pierced ears, a salt-and-pepper goatee, a great smile, a silver and gold Mexican cuff around his wrist, and possibly the most beautiful hands I've ever seen on a human being. He could be a hand model. The only other hands I've seen like those—long, tapered, expressive fingers; short, perfectly groomed, U-shaped nails; and supple, buttery smooth, hairless skin—are on Michelangelo and Leonardo da Vinci Jesuses. Actually, Krunchy looks like Jesus—with a braided ponytail. He's the Krunchy Granola Jesus. From Queens.

Speaking of deities, God gifted Krunchy with divine massage technique. And I *know* my massage like I *know* my red lipsticks. I'm a connoisseur of touch. Don't even say it. I'm not talking about *that* kind of touch. I'm talking about touching that transcends sex. It's a different hue of high. Besides, Krunchy's married. So despite our instant affinity and occasional and innocent flirtiness, he's not a romantic object. It's all part of the whole Krunchy's World experience.

Just being in a dim, comfortable, private room without my clothes on; hearing Stan Getz's tenor sax play a slinky, sinuous, meditative bossa nova over the speakers; and laying my stiff, sore body facedown on Krunchy's table over strategically placed, supportive chest, stomach, and leg ergonomic memory foam cushions—it all made me melt. I felt so much better already and nothing had happened yet! Krunchy had taken extensive notes about the history of my sorry condition before my massage began:

> Just finished writing a book. Very stressed. Tight across shoulders and down the arms. Also stiff neck and sacral-lumbar tension. Needs work especially on muscles in upper back, including trapezius, rhomboids, levator scapulae. Pressure point work to levators and rhomboids at scapulae, scapulae release. She says, "Please help me. Everything hurts."

Krunchy'd determined what I needed was deep Swedish massage with aromatherapy. Groovy, man. I can dig it! I covered my derrière with a freshly laundered sheet and lowered my face into the soft-paper-lined memory foam face cradle. The linens smelled clean and pure. Or maybe it was the room. Either way, aaah. It feels good to feel safe and nurtured and to take care of yourself. I hadn't felt like that for years.

Krunchy knocked on the door. "Ya decent?" he asked.

It made me laugh. "Very *in*decent," I replied.

He came in, slipped his long ponytail down the inside of the back of his Hot Tuna T-shirt, and fastened an oil bottle holster around the waist of his olive fatigue cargo pants. He pulled the sheet up to my neck and placed outstretched palms across my back, pressing them down on me so we could each get the feel of the other. He repeated this all the way to my shoulders and down my arms to my hands, where he stopped, holding them. He moved back to my back, pressing his way down my spine to my lower back, to the backs of my thighs and calves, to the soles of my feet. He held my feet in his hands. Aaah. He rolled the sheet down to my hips. He squirted sweet almond oil into his hands. This essential Indian oil contains ayurvedic herbs with exotic names— ashwagandha, brahmi, shatavari—and familiar ones—rose, rosemary, sage, basil, and cinnamon. Cinnamon warms things up.

Now we were skin to skin. Krunchy laid his artistic and magical hands on me, slowly stroking my spine with long, flowing movements.

"Your back is like granite," he said quietly. "What've ya done to yaself?"

I burst into tears. It was abrupt, totally unexpected, and really embarrassing. It made me feel more naked than my nudity. It had been a long time since anybody touched me for *me*, without expecting something in return. I don't mean money, like Krunchy's fee. I mean, to do something for *me*. I thought about this little pink raincoat I'd worked so hard to get, searched so hard for all across the country, just to please the man I thought was going to marry me. I had so much hope, so much love! But it didn't work. The raincoat didn't make him want to marry me. And just thinking about it made me feel so sad and stupid, the idea that a *raincoat* would make somebody love you enough to want to marry you. And suddenly I realized I've spent my whole life doing this with other clothes and with jewelry and makeup and perfume and shoes for other men I thought I loved whom I wanted to love me. And I just kept doing it and they just kept leaving and the more I thought about it all, the harder I cried. Billy never understood it. I'd always tell my friend, "That's because you're a gay man who really shouldn't be gay. Otherwise you would understand it." And Billy'd say, "Trying to get a man with a raincoat is like trying to catch a fish with a lawn chair."

"Here, Smoothie," Krunchy murmured, handing me a box of tissues.

I looked up at him. He was kind of fuzzy because I was contact lens-less *and* eyeglass-less.

"Thanks," I sobbed. I propped myself up on my elbows and blew my nose.

"*De nada,*" he said. Fuzz apart, I could tell he was smiling. "Got something else for ya." He handed me a palm-sized plastic bag filled with gel and what looked like a dime. It was called The Heat Solution Zap Pac. I held it up to my face to see. "See that metal button in there? Pinch it, like you're trying to break it." As I did, the gel fizzed into a white cloud and got warm. It was really neat. In five seconds, the whole thing had fizzed itself into 130 degrees (Krunchy told me so), and would stay that way for thirty minutes.

"This is so cool!" I said.

"Yep," he said. "Cool and hot." He took the Pac out of my hands and laid it on the small of my back. Aaah. "These are quicker than stones, which ya have to heat in special cookers. Zap Pacs are a more involved way of adding heat and pressure to the massage. Feel like popping some more?"

"Yeah!" I'd forgotten aaall about my crying. Krunchy gave me four more: two for the soles of my feet—aaah-aaah—and two for the palms of my hands—aaah-aaah.

"Okay," he said. "Get back down, Smoothie. And hold on tight to your little buddies." I squeezed the Zap Pacs in my hands. Krunchy gently pushed, wiggled, kneaded, vibrated, rolled, tapped, compressed, and stroked me all over with his hands and arms.

"Am I the worst you've ever seen?" I asked.

"For sure, you're in bad shape," he said, leaning in and using his thumbs to push in and out toward my glutes. Aaah. "But no, you're not the worst I ever saw. You don't get that honor."

"Thank the Lordy."

"Keep breathing, Smoothie. Release it. Let yourself get lost."

"How come you don't mind that I'm such a wreck?" I asked. "I'm just so . . ."

"'Cause underneath it all we're just muscle and bone," he said. "We're all born with the same stuff. And what we do over our lifetime determines what you'll find on the table. Massage is visceral and spiritual."

"No, I mean, you act like you like me."

"I do."

"Why?"

"You've got a really nice personality, a fun way of expressing yaself. And ya don't take things so seriously. Ya have that, what is it, *joie de vivre*. And that comes through. Like meets like."

"I think you're amazing," I mumbled. "*Sui generis*."

"And you're a published author. I think *that's* amazing."

"Aaah, what is *that*? That is just unbelieeevable."

"This?"

"Mm-hm, that."

"It's called a sacral clear-off. It takes all the tension out of your lower back."

"I looove it. *Cleeear-off*."

"Like a clearance sale."

At the sound of those words I felt a rush of happiness spurt up through me like a fabulous fountain, like Anita Ekberg in *La Dolce Vita*, womanly and gorgeous and crazy in a beautiful black evening gown, splashing around in the Trevi Fountain at dawn.

I felt free.

The next day I ran over to my favorite Italian's to tell him all about my religious experience with The Other Jesus.

"I can't believe it," I told Billy, as he made coffee and handed me an ashtray. "It was the first time a man had me between the sheets—solo—and I felt sensual and whole."

"That's great, honey. Your tension is his business. All this guy cares about is that you had a shitty week!"

"Exactly! And he sees me at my worst and he still is nice to me."

"Honey, he likes what I like. He sees *you*. Your mind, your sense of humor, your personality, how you express yourself."

"Yeah, and he keeps his clothes on and I can talk to him about my life. About anything. And he just sort of takes it all in. I can be myself and he's not gay and I'm not sleeping with him. Oh my God. Is it possible that he's my . . . *friend*? I've never said that about any man. Ever. You know how when you call your boyfriend your best friend? I've never said that. God. Krunchy's my *friend*. This has never happened to me!"

"He's making you feel good about yourself," Billy said, pouring the coffee into two big mugs and walking me into the living room. "That's terrific. Plus you gotta love the attention."

"Oh my God, dear. He does this thing. He puts his fists like this, on both sides of your neck? And then he, like, kneads inward? It's like, it's like, like . . . moving human bookends and your head is the book!"

"Something happens to your physiology," Billy said, firing up a Marlboro Light. "You just glow when you're aware someone finds you attractive."

"Yeah, but this is platonic," I said. "We're not dating. We're nondating!"

"Doesn't matter. You look better to yourself. Dilation of pupils . . . something's coming through your skin. When you're depressed, your color's off. It's like after going for a run. You feel flushed, you feel good, you look better."

"Krunchy calls massage 'a passive workout.' He said I still have to do the yoga, though, 'cause one's not a substitute for the other. They're *complementary*. It's a flexibility issue. The body's more accepting of the massage. That's how he can tell whether or not you've done the yoga. It's a sign."

"He sounds like a wonderful masseur," Billy said. "Open, loving, generous."

I sipped my coffee and nodded. "I'm seeing him again next week. I think I may be addicted. Krunchy's my new drug. Oh, and you should have seen me last night, Billy. No makeup, wet hair, yucky nothing outfit, and *glasses* 'cause the contacts were bothering me. They always do in the winter, when there's no humidity. And Krunchy's no Brad Pitt but, you know."

"Suddenly you're naked, even with your clothes on."

"Yeah. Just the Gladiator, that's all I wore. Oh, which reminds me. I've got to get some more. I'm down to two tubes."

"See? You still love those girly things. They're still important to you. They still attract men. You're still who you are. It's just that you feel good with or without them when you're with Brunchy."

"KRUNCHY."

"You've just had so many men who treated you like shit or were unavailable or treated you like an accessory. Brunchy's not the type you'd marry, but your fear about feeling naked without 'the right stuff,' it's . . . it's, like, 'Why do I have these things in the first place?' Is it about impressing other people?"

"I don't know," I said. "Is it?"

"NO!"

"Oh. Okay. Whew!"

"You love fashion and it's fun for you."

"Exactly!"

"But skin and hair are beautiful, too. Your bare skin and your hair are beautiful."

"Oh my God, at the end of the massage, when you're on your back? Krunchy does this thing where he clutches fistfuls of your hair and he just kind of, like, pulls it away from your scalp? Oh. My. God. I'm in his clutches."

"See, Jane? You don't *need* to be cha-cha to look beautiful and turn a Hot Tuna head with a ponytail and glasses into Tarzan."

Maybe not. But I'll *always* need more Gladiator.

A few weeks later, Billy took me out for dinner to The Cheesecake Factory to belatedly celebrate me finishing my book. The restaurant's in Riverside Square, a swanky mall near my place. After polishing off a meal that could—as one Italian character put it in *Moonstruck*—choke a pig (buffalo wings, avocado eggrolls, Factory Burrito Grande, Chicken Madeira, wine, bread, Snickers Bar Chunks and Cheesecake, coffee), we waddled over to the Chanel counter at Bloomies so I could stock up on Gladiator.

"Oh, look," I said. "You know what? Before we get the lipstick, I need toner."

"What kind of printer do you have?" Billy asked.

"TONER!" I yelled. "The kind for your FACE, not your peripherals!" I studied the four different ones: Soothing, Energizing

Radiance, Gentle Hydrating, Oil-Controlling Purifying. Hmmm. "Billy, do you think I need moisture and radiance or hydration and a healthy glow?"

"Can I have a pen?"

"Why?"

"Because that's the stupidest question I ever heard!"

"May I help you?" asked a saleslady.

"Yes," I said, frowning at Billy, who was busy examining his pores in a magnifying mirror and bellowing, "OH MY GOD, THEY'RE THE SIZE OF THE GRAND CANYON! IS IT THE GREEN LIGHTING IN HERE? IT'S HORRIBLE!"

"They let my sick brother out from the institution a little early," I told the startled lady. "He's harmless. Really. Let's not encourage him. Anyway, I need the Energizing Radiance Lotion—"

"Okay," she said.

"And I need all the Gladiator lipsticks you have."

"Gladiator?" she asked.

"Yep," I said. "I want all of 'em."

"Chanel discontinued that color months ago," she said.

"They WHAT?" I screamed.

"They let my twisted sister out from the institution a little too early," Billy said.

"I'm sorry," she said, by now completely confused. "I don't have a single Gladiator."

"But, but that's IMPOSSIBLE!" I screamed. "It just can't BE! What am I gonna DO? My friend Krunchy loves it on me and I love it on me and so does Billy and I've always worn it and—"

"Calm DOWN, Jane," Billy told me. "At least you'll be all energized and radiated for your next massage with Tarzan the vegan. Hahaha."

I called virtually every department store in the entire United States that sells Chanel makeup. Not one had Gladiator. While I'd been engrossed in writing my book, Chanel had quietly conspired to FREAK ME OUT. Okay, try not to panic. Maybe it's not too late. And this is nowhere as bad as the time you lost your makeup bag in Raleigh. What did Krunchy teach you? *Keep breathing. Release it. Let yourself get lost. Aaah.* I opened a TaB, lit a Parliament, and went online. I cruised. Every. Last. Relevant. Viable. Site. No. Gladiator. (Worst-case scenario: I knew I could go to *www.threecustom.com*, where Trae, Scott, and Chad, the Three Custom Color Specialists, can reproduce any lipstick shade known to woman—or gay man—for $50 for two tubes.)

But! I'm a trained reporter. I know how to find information. I was determined to conquer the American Gladiator shortage. And then it occurred to me. Our neighbor to the north! O Canada! I searched the Internet and came upon Andrew's, a cool boutique in Toronto that sells Chanel. I called and a lovely woman named Chisa informed me that she had the very last Gladiator in all of North America! Canadians get their Chanel directly from France. So it was the French connection! I would've bought twenty had Chisa had 'em, but I settled for the one, and we charged it to my battered Visa: $38.35 for a $23.50 lipstick. When I received the package and tore it open, I discovered not only my Gladiator but also all kinds of wonderful Chanel skin product samples and a note from Chisa promising to try and locate more Gladiators and keep in touch.

I love Canada.

My second session with Krunchy was, if this is even possible, more aaah-some than the first. Afterward, I drifted out like a stoner chick. Sitting behind the front desk, Krunchy looked unhappy.

"What happened to you?" I said. "Did Jerry Garcia die?"

"That was back in '95, Smoothie."

"Oh. Well, are you having some kind of horrible mushroom reflux flashback? The bad batch of '73?"

"No, that one was a bumper crop. Very floaty and nice and calm and trippy. It was the '75 crop that sucked. Harsh as LSD."

"Not that you would know or have any personal experience."

"No, of course not. Ah, I'm just bummed because my next client canceled while you were getting dressed."

"Hey, Gigi," Sue said, coming out of another room. She's one of Krunchy's therapists. She has sapphire-tinted hair and teaches yoga on the side.

"Hi, Sue," I said.

"Hi, Gigi," said Shirley, coming out of the bathroom. She's a therapist, too. She's a Puerto Rican redhead who likes singing karaoke and is mad because at thirty she's too old to try out for *American Idol.* She has a really beautiful voice, though. I overheard her singing Alicia Keys's song "Fallin' " a cappella in the parking lot the first night I went to Krunchy's, and for a minute there I thought it was a car radio or a CD playing.

"Hey, Shirley," I said.

"I guess I'm outta here, kids," Krunchy said, slipping on his tan suede shoes. (Almost none of the back in touch therapists are shod while working.)

"Oh, Lands' End," I said. "The 'All Weather Moc.' "

"Nope, these are GBX," Krunchy said. " 'There ought to be a law against shoes this cool'—that's their motto. Kohl's finest. $19.95 on sale. Lands' End's are $29.50."

"That is *very* Jewish of you, Kruncholio. I had no idea."

"Hey, I grew up in New Yawk," he said. "So, Smoothie, want to grab a bite? I think Bischoff's may still be open."

"Sure," I said. "I could handle being force-fed homemade ice cream."

Outside, Krunchy found a stick on the ground and scratched our initials into the freshly laid pavement in front of his shop.

"What's that for?" I said, laughing.

"I declare myself dictator for life of the Gigi fan club," he said.

Wow.

We sat in a booth at Bischoff's and ordered two coffees, and a huge ice water for me. Massage makes you thirsty. Thirsty and cold. Those are good signs. Means all the toxins are out of you, or something. Krunchy opened his wallet to extract one baby picture and a current one of sixteen-year-old Aja, his only child.

"You guys named your daughter after a Steely Dan record?" I asked.

"Record and song," he said, glancing at the menu. "Ya like pecan pie? Warmed?"

"Yeah. I do. Let's share something."

"Okay. Ya like . . . Almond Joy ice cream?"

"Um, I don't know; I've never had it. But it can't be bad, right?"

It was the most delicious ice cream I'd ever tasted. A deeply vanilla-y vanilla bean, with thick strands of sweetened coconut flakes and broken pieces of Almond Joys. Together with the pie, it was aaah.

"I got something for ya," Krunchy said, digging into his satchel. "I made it last night. The Stan Getz CD ya liked? Ya remarked on it."

"Oh my God, that is so sweet! Thank you! Oh. I have something for you, too." I rummaged through my huge tote bag.

"Nothing perishable in there?" he said.

"Only my heart."

"I'll try and keep that in mind."

I withdrew an envelope and gave it to him. It contained an article from *Vogue* about how in the future human massage will be obsolete because machines will be able to do the job. It was illustrated with a photograph of a model lying on what appeared to be a metal assembly line.

"Mm, skin and bones," Krunchy said in a disapproving tone. He shook his head. "She looks like a mannequin." Considering I was busy putting away five gazillion calories, that was exactly the right thing to say.

"So I was telling my friend Billy how stress is your friend," I said.

"Yeah. Stress is my friend. Or, not a friend, but sort of a co-worker. It's the antithesis thing going on. The antidote to civilization is massage."

"You mean because the world can be so bad?"

"Yeah. I've come to believe in the karmic effect of my actions. Everything is all connected and I'm coming to that realization, that there's a vast ocean of interconnectedness. Like the body. I grew up reading science fiction. People with fantastic powers of mind and wristwatch radios."

"Dick Tracy?"

"Yeah. I read this book by Michael Talbot called *The Holographic Universe*. And I thought, Holy shit, that's how I feel. Oh, wow. All these things click into place, how it's supposed to be. How can it be normal for people to have hatred and violence? Light and dark forces. But hey, yin and yang says it all."

"So does ice cream and pie. Are you gonna eat that last bite?"

"It's yours, Smoothie."

"Thank you."

"So the path I want to follow, to help people? It's a big responsibility."

"You changed MY life," I said. "This ice cream! Holy smokes!"

"The only clients who drive me nuts are excessively hairy people—"

"Not my issue."

"No, that's why you're called Smoothie."

"I know."

"People who talk on their cell phones during the massage—"

"You're KIDDING."

"And people who tell you what to do."

"They can't give up control. Type A."

"Right. So, um, about giving up control?"

"Yes?"

"Your lipstick."

"I will NEVER give up my lipstick. NEVER. It's called Gladiator, you know. 'What we do in life echoes in eternity.' I don't know *what* happened to Russell Crowe. But I do know that I own the last tube of Gladiator the Lipstick in all of North America."

"It suits ya. It does."

"So what's the problem?"

"My laundry service tells me they can't get your lips' imprint out of the sheets. Ya don't need Gladiator to come to massage."

"But then I'll be really naked," I said.

"Is that so bad, Smoothie?"

"Well, no. But it's kind of scary. But if you insist . . ."

"Thank you."

I kissed Krunchy's cheek with un-Gladiated lips. I'm sure I left an impression. But after all the talk and ice cream and pie and coffee and water, it was the kind you had to feel, not see.

ACKNOWLEDGMENTS

You know those authors who hole themselves up in cabins in the woods without electricity à la Thoreau and don't talk to anybody for a year while they're writing their books? Me neither. That is *so* not me! Writing a book is a solitary act but it also is not. It takes several significant people to keep you grounded throughout the process and, at the same time, help you fly.

I want to begin by thanking men. You guys kill me. You really do. I don't understand you at all (and I guess it's mutual), but I have to say, you've given me great material over the years. Which has led to a little pink raincoat. Which has led to a book about a little pink raincoat. And about men.

Next are the friends I'm so lucky and honored to have in my life; special friends who happen to be *very* well dressed *and* drop-dead brilliant editors:

Peggy Hackman: THE friend and editor of a lifetime.

Jim "Jaime" M. Naughton: Ditto.

Steve Padilla: *También* ditto.

In alphabetical order, thank you, thank you, too, to:

Lilly Anders, Judith Bachman, Katherine Beitner, Kimberly Berg, Christopher *"Mi Abogado por Siempre"* Bolen, Katherine "Kat" Bomer, Joan Breiter, Elizabeth DeVita-Raeburn, Michelle Dominguez, Bill Ervolino, Annie Ford-Doyle, Shanita Furjanic, Jonathan "Johnny" Gordon, Maryellen Gordon, Mary Hadar, Allen Heinlein, Macarena Hernández, Michael Hill, Bruce Kahn,

Robert Kelley, Tammy Leopold, Elizabeth Llorente, Ana Menendez, Gwen Mitchell, Beth Neelman-Silfin, Emanuel Ory, Manny Roman, Dana Rosen, Bill Smart, Mitch Tuchman, Sharyn "La Shar" Vane, and Eman "Brastrap Petunia" Varoqua.

And a special and very affectionate thank you to:

My excellent, elegant, intelligent, and very stylish agents, Jane Dystel and Miriam Goderich.

My Rayo editor and publisher, Rene Alegria.

My talented book designer Janet M. Evans-Scanlon.

My brilliant senior production editor, Marina "Ass Saver" Padakis.

My excellent director of publicity, Shelby, "La Mensch" Meizlik.

My (gone but not forgotten) marketing director, Elana Stein.

The memory of my beloved friend, Joe McClellan, aka Tío Pepe, who loved and believed in me forever.

The memory of Dr. Marvin Adland, world's best psychiatrist and humanist, who never gave up on me and who once told me, "To be able to write is a wonderful gift—one to be treasured and nurtured."

My mother, Ana Anders, the original (and my favorite) cover girl.

The memory of my handsome, elegant, kind, sweet, and good father, David Anders. *Te quiero mucho, Papi.*

ALSO BY
GIGI ANDERS

BE PRETTY, GET MARRIED, AND ALWAYS DRINK TAB

A Memoir

ISBN 978-0-06-056370-7 (paperback)

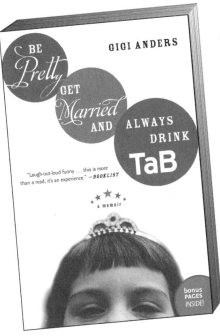

According to her colorful Mami Dearest, the life of young Gigi Anders will be simple if she can remember three maxims—be pretty, get married, and always drink TaB. Thus begins her instruction in the art of being a lady and the side effects of falling in love. As the granddaughter of Eastern European and Russian shtetl-reared grandparents who immigrated as teenagers in the early 1920s to the fierce tropical beauty of Cuba, Anders is heir apparent to a legacy of transatlantic alienation. With dazzling wit and hilarity mined from the depths of loss and yearning, Anders chronicles her journey from beach baby to ostracized exile to vibrant intellectual, along the way balancing her obsession with killer outfits and zaftig, orgasmic meals—always with a can of TaB!—with the more serious pursuits of love, sanity, and lipstick in perfect siren red.